Miles of Tiles

STUDENT MATHEMATICAL LIBRARY
Volume I

Miles of Tiles

Charles Radin

AMERICAN MATHEMATICAL SOCIETY

Editorial Board

1991 *Mathematics Subject Classification*. Primary 52C22;
Secondary 58F11, 47A35, 82D25, 20H15.

The writing of this book was supported in part
by Texas grant ARP 003658-152 and NSF grant DMS 9531584.

ABSTRACT. This is an interdisciplinary book aimed at senior level undergraduate mathematics students. The common thread throughout the book is a type of tiling, of which the best known example is the "kite and dart" tiling of the plane. The book attempts to show that to understand this new type of structure it has been necessary to employ an unusual variety of specialties from mathematics and physics, and that this effort has resulted in new mathematics.

Library of Congress Cataloging-in-Publication Data

Radin, Charles, 1945–
    Miles of tiles / Charles Radin.
        p.   cm. — (Student mathematical library ; v. 1)
    Includes bibliographical references and index.
    ISBN 0-8218-1933-X (softcover : alk. paper)
    1. Tiling (Mathematics) I. Title. II. Series.
QA166.8.R33   1999
516–dc21
                                                                                                                        99-20662
                                                                                                                        CIP

With love to Diane and Sarah

Contents

List of Figures

Preface

"In the world of human thought generally, and in physical science particularly, the most important and most fruitful concepts are those to which it is impossible to attach a well-defined meaning" – an intriguing idea from the physicist H.A. Kramers [Kra]. Even within mathematics, where the approach to knowledge is somewhat different from the physical sciences, a version of this thesis would appeal to those engaged in research: in prosaic terms – anything well-understood is less promising as a research topic than something not-well-understood.

This book is motivated by something not-well-understood, a class of structures exemplified by the "Penrose tilings", or, more specifically, the "kite & dart tilings" (Fig. 1, on p. 2). These structures are not-well-understood on a grand scale, having had significant impact in physics and mathematics, and originating from work done in the 1960's in philosophy! A few decades is a short time in mathematics, so it is reasonable that such a fertile subject is not yet well-understood. Yet it is the subject of this book.

Such tilings differ in a variety of ways from any studied before, and it may be many years before we are confident of how they fit into the body of mathematics, and the most useful ways to view them. Our search to understand them will draw us into many parts of mathematics – including ergodic theory, functional analysis, group representations and ring theory, as well as parts of statistical physics and crystallography. Such breadth invites an ususual format for this mathematics text; rather than present a full introduction to some corner of mathematics, in this book we try to display the value (and joy!) of starting from a mathematically amorphous problem and combining ideas from diverse sources to produce new and significant mathematics – mathematics unforeseen from the motivating problem.

The background assumed of the reader is that commonly offered an undergraduate mathematics major in the US, together with the curiosity to delve into new subjects, to readjust to a variety of viewpoints. The book is self-contained; subjects such as ergodic theory and statistical mechanics are introduced, *ab initio*, but only to the extent needed to absorb the desired insight.

I hope I have imparted in the text the excitement I have enjoyed in the journey through the diverse subject matter. Part of the pleasure has been learning from friends and colleagues. My formal training was in physics, and it would be impossible to name all to whom I am indebted for that; as for mathematics, it is a great pleasure to acknowledge: Persi Diaconis, Charles Fefferman, Richard Kadison, Raphael Robinson, Hao Wang, and especially John Conway and Mark Kac, for insight and inspiration you cannot find in books.

It is also a great pleasure to thank Marjorie Senechal and Jeffrey Lagarias for a great number of useful comments and encouragement; this book would certainly not have appeared without their help. And finally, special thanks are due Stuart Levy for creating the program "subst" for the display of substitution tilings; in particular I used it to make the figure of the quaquaversal tiling used in this book.

Most of this book was written in 1997; for further results see the papers listed in http://www.ma.utexas.edu/users/radin/.

Introduction

The "kite & dart tiling", pictured in Fig. 1, has been widely publicized in the last decade or two. *Why?* There are lots of reasons actually, and in this book we will concentrate specifically on the mathematics which they have inspired, scattered in totally unexpected directions.

The story began in the early 1960's, with the philosopher Hao Wang modeling certain problems in logic [Wan]. It slowly evolved into geometry, in large part from influential work of Raphael Robinson [Rob] and the kite & dart tiling of Roger Penrose [Gar]. We will pick up the story there, and follow it through the twists and turns it has undergone, to the new mathematics that is emerging. It has been a highly interdisciplinary journey, and though we will strongly emphasize the mathematics (chiefly geometry and modern analysis), we will not flinch from analyzing those ideas in physics and crystallography which will help us understand the mathematics. Personally, I find that a good part of the fun of the subject.

Now patterns like Fig. 1 are pretty, but at least once in a while it is useful to face the essence of our endeavor in its raw form. The feature of Fig. 1 that we first emphasize is the *large number* of polygons in it. (There are infinitely many in the full tiling of course.) Analogous

Figure 1. A Penrose "kite & dart" tiling

physical patterns are: a quartz rock made of many atoms; a snail
made of many cells; or a beach made of many grains of sand. In
general, we will be analyzing "global" structures made out of many
small components.

We concentrate not on the external shape of the global structure,
such as the facets of a quartz rock, but on the pattern made at a
much smaller scale by the small components. For a snail this is quite
complicated: the cells gather together into intermediate-size structures which we call organs. For beaches the small scale structure is
rather "random". But rocks are neither as random as beaches, nor
as exotic as snails; the atoms in rocks form patterns of intermediate
complexity, called crystals.

We know roughly why atoms in rocks form crystals while sand grains on the beach or cells in a snail do not; what we say is that *there are different laws governing the production of these structures* (laws studied in physics, biology etc.), and these different laws naturally lead to different results. But this is all a bit vague, and when one examines this explanation carefully there are serious but very interesting difficulties with it. This will be our subject – why certain kinds of laws or rules seem to produce very special global structures, such as quartz crystals. (Snails are a much harder problem, and beaches are too simple; neither will be mentioned seriously again!) In the next few paragraphs we must get more specific about this idea of rules which produce structures.

This book is about mathematics, not physics, so it will be useful to have in mind models of this structural phenomenon other than rocks, with all their irrelevant details. The general model we will use is the jigsaw puzzle, one made with *very* many pieces with bumpy edges, pieces which we will call "tiles". If one imagines the tiles to have various colors painted on them it is easy to see how a tile contributes to a global structure or pattern. We will ignore any such colors on the tiles, and only concentrate on their shapes – as if we turned the jigsaw puzzle pieces over; we think of the global pattern as "consisting of" these special shapes fitted together. The bumpy edges of the tiles play an important role in determining how the tiles are allowed to fit together to form the global pattern. (We will also consider 3-dimensional versions of the more traditional 2-dimensional puzzles.)

But rocks will still be useful to us; we will use features of rocks to guide us in our mathematical analysis of patterns. For instance, the tiles in a jigsaw puzzle could represent *any* structure if there were no restrictions. If you wanted certain star-shaped pieces to lie in particular places in a puzzle, you could just chop up the intervening space into other pieces to accommodate the stars. Now one of the reasons the atoms in rocks do not appear in arbitrary (local) structures is that there are only a small number of possible components, the 92 different naturally occurring kinds of atoms. So we ask: if you were a manufacturer of jigsaw puzzles and were limited to using, say, 92 different tiles of your choice (but could make as many copies of each shape as you wanted), what kind of giant patterns could you produce? Remember that for us the pattern has nothing to do with colored pictures on the tiles, but merely with the manner in which the various tiles fit together. With this constraint of only 92 different shapes to use it is no longer clear that you could make *any* (local) pattern of tiles, and the question is: what kinds of patterns could you make? This is where our discussion about rules of production has led us; we assume we have some small number of different kinds of elementary building blocks, and ask what kinds of patterns can be made out of them given the restriction that the pieces have to fit together like a jigsaw puzzle. Later we will discuss why the restriction that the pieces fit together should be thought of as a law of production.

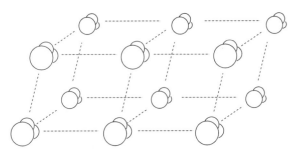

Figure 2. 12 unit cells of a periodic configuration

Now let's refine what we mean by "kinds of patterns". Again we look at crystals for inspiration. The distinguishing special feature of crystals is that they each have a unit cell from which the global pattern is obtained by translation, as in Fig. 2. (The dashes in the figure show the translations separating the 12 unit cells, each unit cell consisting of three spheres.) This repetition is an example of an "order" property, very different from the randomness of the positions of the sand grains of a beach. Also, periodic structures such as crystals are limited in the symmetries they may have; for instance, a crystal can only have axes with 2, 3, 4 and 6-fold rotational symmetry [HiC; p. 84]. So in studying the "kinds" of structures, we will concentrate on the order and, especially, the *symmetry* properties they may possess.

It is time to examine some examples. First consider the periodic jigsaw using the following K^2 different shapes, where $K \geq 2$ is fixed but arbitrary. They are all basically unit squares, but with certain bumps coming out of some edges and dents going into some edges (any bump fitting into any dent), following the formula: the tile labeled (i, j) (where $1 \leq i, j \leq K$) has i dents on its top edge, $(i + 1)$ bumps on its bottom edge, j dents on its left edge and $(j + 1)$ bumps on its right edge. The exceptions are: $(1, j)$ has K dents on its top edge; $(i, 1)$ has K dents on its left edge; (K, j) has 1 bump on its bottom edge; and (i, K) has K bumps on its right edge. Fig. 3 shows tile $(3, 5)$, assuming $5 < K$. It's not hard to show that the only way to put these together into a big jigsaw puzzle makes a "$K \times K$ unit cell" (several are shown in Fig. 4 for $K = 6$), and builds a periodic pattern from copies of the cell. Notice that although the specification of the unit cell is not unique (see an alternative choice in Fig. 5), in a sense the full pattern is unique: there is really only one way to build a global structure from these tiles, up to an overall rigid motion of the plane. (See Appendix I for a review of congruences of Euclidean 2- and 3-dimensional space.)

Figure 3. The tile (3,5)

1,1	1,2	1,3	1,4	1,5	1,6	1,1	1,2	1,3	1,4	1,5	1,6
2,1	2,2	2,3	2,4	2,5	2,6	2,1	2,2	2,3	2,4	2,5	2,6
3,1	3,2	3,3	3,4	3,5	3,6	3,1	3,2	3,3	3,4	3,5	3,6
4,1	4,2	4,3	4,4	4,5	4,6	4,1	4,2	4,3	4,4	4,5	4,6
5,1	5,2	5,3	5,4	5,5	5,6	5,1	5,2	5,3	5,4	5,5	5,6
6,1	6,2	6,3	6,4	6,5	6,6	6,1	6,2	6,3	6,4	6,5	6,6
1,1	1,2	1,3	1,4	1,5	1,6	1,1	1,2	1,3	1,4	1,5	1,6
2,1	2,2	2,3	2,4	2,5	2,6	2,1	2,2	2,3	2,4	2,5	2,6
3,1	3,2	3,3	3,4	3,5	3,6	3,1	3,2	3,3	3,4	3,5	3,6
4,1	4,2	4,3	4,4	4,5	4,6	4,1	4,2	4,3	4,4	4,5	4,6
5,1	5,2	5,3	5,4	5,5	5,6	5,1	5,2	5,3	5,4	5,5	5,6
6,1	6,2	6,3	6,4	6,5	6,6	6,1	6,2	6,3	6,4	6,5	6,6
1,1	1,2	1,3	1,4	1,5	1,6	1,1	1,2	1,3	1,4	1,5	1,6
2,1	2,2	2,3	2,4	2,5	2,6	2,1	2,2	2,3	2,4	2,5	2,6

Figure 4. Part of a periodic tiling, with unit cells outlined

Next consider some tilings, due to Jarkko Kari and Karel Culik, made from copies of the shapes in Fig. 6. As is shown in [Cul], one can build arbitrarily large collections of these tiles, but none with a unit

1,1	1,2	1,3	1,4	1,5	1,6	1,1	1,2	1,3	1,4	1,5	1,6
2,1	2,2	2,3	2,4	2,5	2,6	2,1	2,2	2,3	2,4	2,5	2,6
3,1	3,2	3,3	3,4	3,5	3,6	3,1	3,2	3,3	3,4	3,5	3,6
4,1	4,2	4,3	4,4	4,5	4,6	4,1	4,2	4,3	4,4	4,5	4,6
5,1	5,2	5,3	5,4	5,5	5,6	5,1	5,2	5,3	5,4	5,5	5,6
6,1	6,2	6,3	6,4	6,5	6,6	6,1	6,2	6,3	6,4	6,5	6,6
1,1	1,2	1,3	1,4	1,5	1,6	1,1	1,2	1,3	1,4	1,5	1,6
2,1	2,2	2,3	2,4	2,5	2,6	2,1	2,2	2,3	2,4	2,5	2,6
3,1	3,2	3,3	3,4	3,5	3,6	3,1	3,2	3,3	3,4	3,5	3,6
4,1	4,2	4,3	4,4	4,5	4,6	4,1	4,2	4,3	4,4	4,5	4,6
5,1	5,2	5,3	5,4	5,5	5,6	5,1	5,2	5,3	5,4	5,5	5,6
6,1	6,2	6,3	6,4	6,5	6,6	6,1	6,2	6,3	6,4	6,5	6,6
1,1	1,2	1,3	1,4	1,5	1,6	1,1	1,2	1,3	1,4	1,5	1,6
2,1	2,2	2,3	2,4	2,5	2,6	2,1	2,2	2,3	2,4	2,5	2,6

Figure 5. Part of a periodic tiling, with different unit cells outlined

cell from which a tiling could be constructed by repeated translation as above.

Instead of talking about arbitrarily large collections of tiles, we will take the plunge and discuss from now on infinite collections which

Figure 6. The Kari-Culik tiles

Figure 7. The "random tile"

fill the whole Euclidean 2- or 3-dimensional space. We will also use the word "tiling" in place of jigsaw puzzle for collections of tiles; by a tiling of a space we just mean a collection of tiles which completely covers the space, and such that for each pair of tiles in the tiling the interiors have empty intersection. As we shall see, there is not really much difference in dealing with infinite rather than arbitrarily large collections of tiles, and it will simplify some discussions. In particular, from the tiles of Fig. 6 one can make uncountably many tilings of the plane no two of which are congruent, but as noted above one cannot make a periodic one, that is, one made up of repeated translations of a unit cell as in Fig. 4.

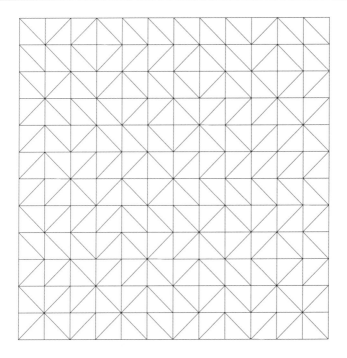

Figure 8. Part of a "random tiling""

We next consider the "random" tile in Fig. 7. The only tilings one can make out of copies of this tile are vaguely like a checkerboard, with the tiles pairing up along their hypotenuses to make the squares of what we will call a "random checkerboard" as in Fig. 8. Note that in pairing up, each square of the checkerboard is filled with a pair of tiles in one of two possible orientations, and that these two possible orientations (think of them as "red" and "black") are independent in different squares. We could thus think of the possible tilings with these tiles as *any* tilings made with (aligned) red and black squares – not just those alternating in color as in a real checkerboard. This means we *can* tile in very complicated ways; but in contrast with the previous example, we can *also* tile periodically, that is, in a very simple way, for instance the usual red-black checkerboard.

We want to contrast one feature of the tiles of Fig. 4 and of Fig. 7. From the first set we found that there was essentially only one tiling that could be made; any two tilings were in fact congruent. From the second set we found we could make a very wide variety of tilings, which had little to do with one another. We noted earlier rules or laws that produce global structures from components, and this is an appropriate place to expand on that.

It would be convenient if our rules took a set of tiles such as those in Fig. 6 and produced a specific tiling. However we are trying to understand how structures such as crystals are made, and the rules of nature are not of this simple type. As we shall see in Chapter 2, given a specification of particles such as iron atoms (or better yet, iron nuclei and electrons), the physical rules or laws that govern the production of bulk iron at low temperature do not actually pick out a specific particle configuration such as a particular crystal; the laws (called statistical mechanics) actually specify a large *collection of particle configurations*. So too the rules we will deal with will associate with a given set of tiles such as Fig. 6 not one specific tiling, but a (large) *collection of tilings*.

For instance, with a set $\mathcal{A}$ of tiles such as Fig. 6 we can associate the set $X_{\mathcal{A}}$ of *all possible* tilings that one could make using congruent copies of those tiles. We think of this as a "rule" for $\mathcal{A}$. (We will consider other types of rules for $\mathcal{A}$ later, which associate special subsets of $X_{\mathcal{A}}$, but unless otherwise indicated, given a set $\mathcal{A}$ of tiles "the" rule for $\mathcal{A}$ associates the set $X_{\mathcal{A}}$ of *all possible* tilings).

Now if one is given a tiling x and wants to determine whether or not it "follows from the rule associated with $\mathcal{A}$", that is, whether or not x is in $X_{\mathcal{A}}$, one must check three things: that the tiles making up x are congruent to elements of $\mathcal{A}$; that in x they leave no uncovered gaps in the plane; and that they never overlap in x. Notice that this process could be carried out by examining x in any fixed disk D of diameter large enough to properly contain any of the tiles of $\mathcal{A}$, and moving this viewing window throughout x; it can be seen whether or not x belongs to $X_{\mathcal{A}}$ by such a *local* examination, where the word "local" emphasizes that at no time do we need to examine a portion of x larger than the finite size of our fixed disk D. In summary: a rule associates with some finite set $\mathcal{A}$ of tiles not one tiling but some *set* X of tilings, and the essence of the rule is what one must do to see if a candidate tiling x satisfies the rule, that is, belongs to X. It is in this way that we have classified rules as local or not.

Getting back to the contrasted feature of the set $\mathcal{A}_4$ of tiles in Fig. 4 and $\mathcal{A}_7$ of the tile in Fig. 7, the difference we noted is that all the tilings in $X_{\mathcal{A}_4}$ are very similar to one another (in fact any two are congruent), whereas there are tilings in $X_{\mathcal{A}_7}$ which are very dissimilar from one another. The set $\mathcal{A}_6$ of tiles shown in Fig. 6 is of a different sort. We said that $X_{\mathcal{A}_6}$ contains tilings which are not congruent to one another. But these tilings are still very similar in a sense slightly weaker than congruence: every finite substructure of any one tiling in $X_{\mathcal{A}_6}$ has congruent copies in any other tiling in $X_{\mathcal{A}_6}$, with the consequence that one cannot tell the difference between the tilings in $X_{\mathcal{A}_6}$ by inspecting only finite portions of them. Now the point is, we consider the cases of $X_{\mathcal{A}_4}$ and $X_{\mathcal{A}_6}$ as satisfactory, but not that of $X_{\mathcal{A}_7}$. (This is partly motivated by the statistical mechanics of solids. After developing some ideas in Chapter 2 we will think of $X_{\mathcal{A}_7}$ as resulting from an "accidental symmetry" which allows both orientations of the pairs of tiles making up squares.) With this prejudice firmly in place, we will concentrate on sets $\mathcal{A}$ of tiles without such accidental symmetries, tiles which only produce sets $X_{\mathcal{A}}$ of tilings in which each pair, while not necessarily congruent, is locally indistinguishable in a sense made precise later. One consequence is that for the sets $\mathcal{A}$ of tiles which we will consider, the tilings in $X_{\mathcal{A}}$ are either all simple (for instance periodic), or all complicated.

Consider next the tiles in Fig. 9 and the following "substitution rule" for making tilings from them. For each tile T we have a way

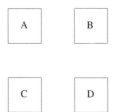

Figure 9. The Morse tiles

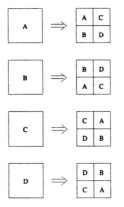

Figure 10. The Morse substitution system

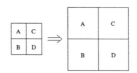

Figure 11. Expansion

(Fig. 10) to associate with it a collection of tiles at a smaller scale (a factor $1/\gamma < 1$ times those in Fig. 9, with $\gamma = 2$ in this case), in a

D	B	B	D	B	D	D	B	B	D	D	B	D	B	B	D
C	A	A	C	A	C	C	A	A	C	C	A	C	A	A	C
C	A	A	C	A	C	C	A	A	C	C	A	C	A	A	C
D	B	B	D	B	D	D	B	B	D	D	B	D	B	B	D
C	A	A	C	A	C	C	A	A	C	C	A	C	A	A	C
D	B	B	D	B	D	D	B	B	D	D	B	D	B	B	D
D	B	B	D	B	D	D	B	B	D	D	B	D	B	B	D
C	A	A	C	A	C	C	A	A	C	C	A	C	A	A	C
C	A	A	C	A	C	C	A	A	C	C	A	C	A	A	C
D	B	B	D	B	D	D	B	B	D	D	B	D	B	B	D
D	B	B	D	B	D	D	B	B	D	D	B	D	B	B	D
C	A	A	C	A	C	C	A	A	C	C	A	C	A	A	C
D	B	B	D	B	D	D	B	B	D	D	B	D	B	B	D
C	A	A	C	A	C	C	A	A	C	C	A	C	A	A	C

Figure 12. A Morse tiling

fixed relation in space to T. What we do then is start with some tile somewhere in the plane, "break it up" as is done to tile A in Fig. 10, then expand the small tiles by the factor $\gamma > 1$ as in Fig. 11. We repeat this for each of the tiles we now have, again and again; see Fig. 12. After infinitely many iterations we have a tiling of the plane, at least if we are careful in choosing each time the place about which to expand. It turns out that different sequences of choices of points about which we expand can lead to tilings which are noncongruent; so a "substitution rule" associates a large number of special tilings with a set $\mathcal{A}$ of tiles.

This is a useful way of producing tilings; one can produce interesting tilings by this substitution method. But in a way it seems too easy; we expect that a more complicated mechanism should be required to govern the construction of the complicated structures of interest to us. One hint to an understanding of this situation is that the substitution rule is not as "local" as was the one used for the tiles of Figs. 4, 6 or 7, because to check if a tiling is correctly made by this substitution process one must examine arbitrarily large regions of the tiling to see if they are put together correctly. We need to discuss this difference, because it goes to the heart of our subject. After all, we said we thought the atoms in quartz make different patterns from the cells of a snail because they follow different rules, so we need to be fussy about what kind of rules we use in making our tilings.

One of the big lessons in physics in this century was the choice made, following Einstein, to follow the field approach to physics of Maxwell's theory of electromagnetism rather than the action-at-a-distance approach of Newton's theory of gravitation [EiI]. In electromagnetism the influences at a given point (and time) are determined by the immediate environment of that point, while in classical gravitation the motion of a planet at one location is influenced (instantaneously) by the position of the sun (and planets) far away. Postponing the details until Chapter 2, we just note here that the laws governing the atomic structure of solids are also of the local variety, and we use this fact to "prefer" such rules in our tiling problems. As we noted before, the rule which, with any given set of tiles, associates *all* tilings that one can make with those tiles, is local, while a rule which associates with a set of tiles only those tilings made by a special technique may or may not be local; for instance a substitution rule is not local. However, substitution rules will still be useful – for instance they will lead to new ideas about the notion of symmetry.

Fundamentally, symmetry means the invariance of something when that something is acted upon in some way. This is so general, and so useful, it is hard to imagine altering it. Indeed, when we say there are new ideas about symmetry, what we are referring to is much more specific. By far the most useful symmetries of patterns

Figure 13. A pinwheel tiling

in space correspond to invariance of the pattern under very special transformations: rigid motions of the underlying space. The patterns we will be dealing with, such as the kite & dart (Fig. 1) or pinwheel (Fig. 13) tilings, are usually not invariant under any nontrivial rigid motion. But the notion of symmetry for such a tiling can be altered by focusing not on the tiling itself but on the relationships between its component tiles.

For each tiling we can construct a set of "frequencies" of its finite parts. That is, for each finite collection p of tiles in the tiling x, count the number of times p appears – in the same orientation – in a ball of volume N, and divide by N, defining, in the limit $N \to \infty$, a frequency $\nu(p)$. (We will deal with the existence of these limits.) One can ask whether such frequencies are invariant when the tiling x is moved by a rigid motion, such as a rotation. The interesting thing about, say, a kite & dart tiling, is that all its frequencies are invariant under a rotation by $2\pi/10$ even though no kite & dart tiling is itself invariant under a rotation by $2\pi/10$.

So the new notion of symmetry is still geometric, in the sense of corresponding to rigid motions; what is new is a shift in the quantity that is invariant – instead of the tiling itself, we focus on the frequencies of its finite parts.

In a nutshell our book will be about how global patterns are produced, that is, the kinds of production rules they have, and the ways in which these global patterns can exhibit order and symmetry. The production rules we consider do not necessarily produce single or unique structures, even up to congruence. But as discussed earlier they almost do this; any two structures produced by such a rule must be indistinguishable by "local" inspection, looking at any finite portion of the structure.

Chapter 1

Ergodic Theory

§1. The idea behind ergodic theory

We are trying to analyze global patterns in space which are produced by "local" rules – rules which only depend on pieces of the pattern of limited size, such as pairs of neighboring tiles – and to determine to what extent the global patterns are orderly or symmetric.

The question arises: what tools are appropriate for the analysis of such patterns? In this chapter we will see how ergodic theory is used to analyze order properties; in Chapter 4 we will extend the same tools to analyze symmetry properties.

As far as order properties are concerned, two of the examples we mentioned in the Introduction represent opposite extremes: the periodic tilings are the most orderly possible, and the random checkerboard tilings are the least orderly possible. Here we are using a meaning of orderliness taken originally from probability theory, then imported into physics, ergodic theory, and other areas. A more descriptive term for this measure of orderliness is "long range order", and before giving a formal definition we note the intuition behind it.

Imagine we can see the finite portion of a d-dimensional tiling near us, and although we know everything about the tiling we don't

know where we are in the tiling in absolute terms. We measure the
degree of long range order of the tiling by the *extent to which we can
infer, from the features near us, features of the tiling far away from
us.* We will use a probabilistic approach to justify such inferences. So
imagine we know that where we are there is some finite collection p
of tiles and we want to measure the degree to which we can expect
that at the location relative to us by the vector $t \in \mathbb{R}^d$ there is a finite
collection q. Consider these two "events", p and $T^t q$ (we assume q
refers to an alternative finite collection at our present location, so T^t
translates it to the desired place; t is the position of $T^t q$ relative to
q). Since we don't know where we are in the tiling, we try to get an
estimate of how likely it is that these two events coexist. We imagine
that any place in the tiling at which the collection p occurs could be
our location, and we treat them all as equally likely. Take a large
ball of volume N centered anywhere in the tiling, count the number
of times the two events p and $T^t q$ occur, and divide by the volume
N. We restrict attention to tilings in which this quantity has a well-
defined limit as $N \to \infty$, independent of the location of the center of
the ball. We call this the "frequency" $\nu(p \cap T^t q)$ of the joint event: p
and $T^t q$, and we use this frequency to measure our expectation that
q is in the tiling at the relative location t to us.

   The usual model for a structure with extreme lack of order is a sequence of independent coin flips, which we have mimicked in our random checkerboard tilings. Note that for these tilings the only translations t for which $\nu(p \cap T^t q)$ could possibly be nonzero are those with
integer coordinates, $t \in \mathbb{Z}^d$. But then, once $t \in \mathbb{Z}^d$ translates q so that
$T^t q$ doesn't overlap p, we have $\nu(p \cap T^t q) = \nu(p)\nu(T^t q) = \nu(p)\nu(q)$.
(From our assumption, it is automatically true that $\nu(T^t q) = \nu(q)$ for
any q and t.) Generalizing slightly from that extreme case, we say a
structure "does not have long range order" if for fixed p and q in it
and large t we have $\nu(p \cap T^t q) \approx \nu(p)\nu(q)$ or, to be more quantitative,
if

$$\lim_{|t| \to \infty} |\nu(p \cap T^t q) - \nu(p)\nu(q)| = 0, \qquad (1.1)$$

where we use the notation $|t| = (t_1^2 + t_2^2 + \cdots + t_d^2)^{1/2}$ for the usual Euclidean length. Alternatively, given the knowledge of p at our location, we have the most knowledge about some q at other locations when p determines the rest of the tiling completely, as is the case for periodic tilings if p contains a unit cell. In such a situation, since $\nu(p \cap T^t q)$ equals either $\nu(p)$ or 0 for any q and t, neither of which equals $\nu(p)\nu(q)$, it follows that $|\nu(p \cap T^t q) - \nu(p)\nu(q)|$ is bounded away from 0 for all t. So when we say a tiling exhibits long range order, that is,

$$\lim_{|t| \to \infty} |\nu(p \cap T^t q) - \nu(p)\nu(q)| \neq 0, \qquad (1.2)$$

we are suggesting a similarity to periodic tilings.

An "intermediate" case that will appear later is the situation where $[\nu(p \cap T^t q) - \nu(p)\nu(q)]$ is small for "most" t; that is,

$$\lim_{N \to \infty} (1/N) \int_{B_N} |\nu(p \cap T^t q) - \nu(p)\nu(q)|\, dt = 0, \qquad (1.3)$$

where B_N is a ball, centered at the origin, of volume N, which only requires $[\nu(p \cap T^t q) - \nu(p)\nu(q)]$ to go to 0 on average, allowing it to stay bounded away from 0 for some t.

In summary, we use the behavior of $|\nu(p \cap T^t q) - \nu(p)\nu(q)|$ for large $|t|$ to characterize a tiling between the extremes of a completely ordered periodic tiling and a completely disordered random checkerboard tiling. (In ergodic theory the situation of (1.1) is called "strong mixing", and that of (1.3) is called "weak mixing".)

The foundation for the subject of ergodic theory is the fact that frequencies such as $\nu(p)$, which we defined by:

$$\nu(p) = \lim_{N \to \infty} \frac{\text{the number of occurrences of } p \text{ in a ball of volume } N}{N},$$
$$(1.4)$$

can be related to other quantities. To discuss this we will need to introduce some extra structure in our mathematics.

§2. Some mathematical structure

We are studying the relationship of small components within a large
structure. One model for that is a very long novel made up of "letters" drawn from some "alphabet" $\mathcal{A}$. For us the letters are polygonal
or polyhedral tiles, and the alphabet is a finite collection of such letters, as in Fig. 6. One difference between our situation and a literary
analogue is that we refer not only to the *relative* geometric relationships between our letters, but to their *absolute* positions. When we
describe a tiling we think of each tile as occupying a specific position
in space. Because a continuum such as the plane or space is more
complicated than a cubic lattice such as $\mathbb{Z}^d$, we will introduce the
necessary mathematical structure in two steps. The first time around
we will only consider "square-like" tiles as in Fig. 6, because the associated tilings have a natural labeling through $\mathbb{Z}^2$. To be more specific
we begin with the following definition.

Definition 1.1. A set of polyhedra in $\mathbb{R}^d$ is called "square-like" if
each possible tiling by these tiles is congruent to one in which the
centers of the tiles are precisely $\mathbb{Z}^d \subset \mathbb{R}^d$ (considered as a Euclidean
space).

For square-like d-dimensional tiles the study of their tilings is
essentially the same as the study of those in which the centers of
the tiles are precisely $\mathbb{Z}^d$, and for simplicity we will mostly be concerned with the special case $d = 2$. It is then natural to associate
a tile with each $j \in \mathbb{Z}^2$, and therefore the tilings, up to congruence, with (certain) points in $\mathcal{A}^{\mathbb{Z}^2}$, which is the set of all assignments
$x \equiv \big\{ \{x_j\} : x_j \in \mathcal{A},\, j \in \mathbb{Z}^2 \big\}$ of a letter $x_j \in \mathcal{A}$ to each site j in the
square lattice $\mathbb{Z}^2 \subset \mathbb{R}^2$, or in other words, as the set of functions on
$\mathbb{Z}^2$ with values in $\mathcal{A}$. (We are using here a common notation generalizing that of sequences, namely we think of $x = \{x_j : m \in \mathbb{Z}^2\}$ as
having coordinates x_j, one for each site $j \in \mathbb{Z}^2$.) Most of what we do
can be generalized easily from $\mathbb{Z}^2$ to $\mathbb{Z}^d$, so we will often refer to this
more general context.

As a set, $\mathcal{A}^{\mathbb{Z}^d}$ has too little structure to bear much analysis, so we will put a metric on it to measure how far apart a pair of tilings is. There is no natural choice, but we will define our metric m with the following in mind. We want two tilings or functions in $\mathcal{A}^{\mathbb{Z}^d}$ to be close in our metric if and only if they are identical at those points of $\mathbb{Z}^d$ which are in a large ball about the origin. This is desirable because it would allow us to focus on local aspects of the tilings. For instance, if we want to compare two tilings x and $\tilde{x}$ near the point $j \in \mathbb{Z}^d$, we first consider the translated tilings $T^{-j}x$ and $T^{-j}\tilde{x}$ which, near the origin, are the same as x and $\tilde{x}$ near $j \in \mathbb{Z}^d$. Then the number $m(T^{-j}x, T^{-j}\tilde{x})$ would be a measure of how much x and $\tilde{x}$ differ near j, since the metric gives little weight to the sum total of all points far away from the origin.

We now construct a metric m with the above features. Starting with the metric $\tilde{m}$ on $\mathcal{A}$ for which $\tilde{m}(a, b) = 1$ for $a \neq b$, we construct the metric m on $\mathcal{A}^{\mathbb{Z}^d}$ by $m(x, y) \equiv \sum_{j \in \mathbb{Z}^d} \tilde{m}(x_j, y_j)/2^{|j|}$. For instance, let's use this metric to compute the distance between the tiling x of Fig. 4 and the tiling y obtained from x by translating x one unit to the right. Note that x and y have different tiles at *every* site of $\mathbb{Z}^2$. So $m(x, y) = \sum_{j \in \mathbb{Z}^2} \tilde{m}(x_j, y_j)/2^{|j|} = \sum_{j \in \mathbb{Z}^2} 1/2^{|j|}$. In fact *every* translate y of x which doesn't coincide perfectly with x differs from x at all sites of $\mathbb{Z}^2$ and therefore is separated from x by this distance. This would not be true in general, for instance if we had taken for x the tiling of Fig. 8 or Fig. 12.

For those familiar with point-set topology, this metric m gives the product topology for this product space $\mathcal{A}^{\mathbb{Z}^d}$. Again, we do not claim this metric is particularly natural in detail, but it does have the properties we wanted.

Also useful is a standard fact (Tychonoff's theorem) that $\mathcal{A}^{\mathbb{Z}^d}$ is compact in this metric, which means that any sequence of tilings of $\mathcal{A}^{\mathbb{Z}^d}$ has a subsequence convergent to a tiling in $\mathcal{A}^{\mathbb{Z}^d}$. (Recall that although our tilings are "unbounded" in $\mathbb{R}^d$, each such tiling is just a point in the space $\mathcal{A}^{\mathbb{Z}^d}$ and we are referring to compactness in this space, not in $\mathbb{R}^d$.)

Now the main reason we are using a description of tilings which emphasizes their positions in space, that is, which distinguishes between a tiling and a translation of that tiling, is that by so doing we can make use of a natural representation on $\mathcal{A}^{\mathbb{Z}^d}$ of the additive group $\mathbb{Z}^d$, namely the map which, for each $t \in \mathbb{Z}^d$, takes $x \in \mathcal{A}^{\mathbb{Z}^d}$ to $T^t x \in \mathcal{A}^{\mathbb{Z}^d}$, where of course $(T^t x)_j \equiv x_{j-t}$ for $j \in \mathbb{Z}^d$. (See Appendix II for a review of some notation and results about group representations.) This is a representation of the group $\mathbb{Z}^d$ since $T^t(T^s x) = T^{t+s} x$ for all $s, t \in \mathbb{Z}^d$, and in fact it is continuous in the sense that $T^t x$ is continuous in x for fixed t. We will refer to T^t as *translation* by $t \in \mathbb{Z}^d$.

Consider the frequencies mentioned in the Introduction. A tiling $x \in \mathcal{A}^{\mathbb{Z}^d}$ is an infinite collection of letters $x_j \in \mathcal{A}$, but it is useful to also consider its finite subcollections, the restrictions of x to finite subsets of $\mathbb{Z}^d$. We naturally call such restrictions "words" in x. The frequencies we refer to are the frequencies $\nu(p)$ of the finite words $p \subset x$. Intuitively they are the fractions

$$\frac{\text{the number of occurrences of } p \text{ in a ball of volume } N}{N}, \qquad (1.5)$$

or more properly, the limits of such fractions as $N \to \infty$. Think of the "occurrences" in the fraction as different translates of the word p; in that way the fraction can be interpreted as something obtained by averaging over translations. Our next goal is to analyze the existence of such limiting averages, and techniques for computing them. This is a fundamental subject, and the main result in it was obtained in 1931 by George David Birkhoff [Wal; p. 34], his pointwise ergodic theorem. This theorem shows a connection between such frequencies, which are averages over spatial translations, and averages over tilings. In principle this is merely a shift of viewpoint; instead of thinking of the fraction as referring to translates of the word p appearing in different places in one fixed tiling, one could reinterpret this as referring to the word p appearing fixed in space as we make appropriate translations of the tiling. This is only "in principle" however, because Birkhoff's theorem is not all that easy to prove, and we won't attempt to do so here.

Now the mathematics of averaging on a compact metric space S such as $\mathcal{A}^{\mathbb{Z}^d}$ is, since the work of Andrei Kolmogorov [Kol], part of the theory of integration of functions on S. We will need little of the details of integration theory, mostly just some of the notation and some basic facts, which we include in Appendix III. As a (very brief) summary we recall that averaging over tilings amounts to associating, with any given function f on the space of tilings, an integral $\mathbb{I}(f)$ of f. So averaging over tilings means integrating functions defined on the set of tilings. And as with the usual integration of functions on the real line, it is appropriate that the integration take into account any symmetry or invariance of the objects we are integrating over. For the real line this corresponds to the "change of variable" formula:

$$\int_{[a,b]} f(x)\,dx = \int_{[a+c,b+c]} f(y-c)\,dy, \qquad (1.6)$$

which one requires for instance so that if one interprets the integral of a positive function as the area under its graph, that area will be invariant under a translation of the graph.

There is a similar requirement we make in our integration over tilings. We are dealing with functions on $\mathcal{A}^{\mathbb{Z}^d}$, and integrals I of such functions. To discuss invariance of our integration it is useful to "lift" the action of translations $t \in \mathbb{Z}^d$ from the set $\mathcal{A}^{\mathbb{Z}^d}$ to the functions and integrals. This means that for any $t \in \mathbb{Z}^d$ and function f on $\mathcal{A}^{\mathbb{Z}^d}$ we define the new function $T^t f$ on $\mathcal{A}^{\mathbb{Z}^d}$ by $T^t f(x) = f(T^{-t}(x))$. It is easy to check that this defines a representation of $\mathbb{Z}^d$ on the space of all complex-valued functions on $\mathcal{A}^{\mathbb{Z}^d}$. And having done that, we can also lift the action of $\mathbb{Z}^d$ to integrals $\mathbb{I}$ by defining new integrals $T^t\mathbb{I}$ through $T^t\mathbb{I}(f) = \mathbb{I}(T^{-t}(f))$. This leads to the definition: An integral $\mathbb{I}$ is called "invariant" if $T^t(\mathbb{I}) = \mathbb{I}$ for all $t \in \mathbb{Z}^d$. Requiring that our integrals over tilings be invariant plays the same role as it does for the usual integration over the real line (think about this), and corresponds to the fact that our averaging arises from the frequencies discussed above, in which we count the number of occurrences in a tiling. Such averaging is at heart invariant under translations. This is the reason invariant integrals are a major ingredient in ergodic theory, as we see in the following version of the fundamental theorem of ergodic theory [Wal; p. 160].

Theorem 1.2. *Suppose there is a continuous representation T of the group $\mathbb{Z}^d$ on the compact metric space X, and $\mathbb{I}$ is an invariant integral on X. Then the following three conditions are equivalent:*

i) $\mathbb{I}$ is the only invariant integral on X;

ii) for every continuous function f on X and $x \in X$,

$$|\frac{1}{N} \sum_{t \in B_N} [T^t f(x) - \mathbb{I}(f)]| \xrightarrow[N \to \infty]{} 0; \tag{1.7}$$

iii) for every continuous function f on X,

$$\sup_{x \in X} |\frac{1}{N} \sum_{t \in B_N} [T^t f(x) - \mathbb{I}(f)]| \xrightarrow[N \to \infty]{} 0. \tag{1.8}$$

To see the connection between this theorem and our discussion of frequencies, consider a continuous function of the type χ_p associated with some word $p \in \mathcal{A}^S \subset \mathcal{A}^{\mathbb{Z}^d}$; that is, χ_p is the indicator function of the cylinder set $C_p \equiv \{x \in X : x_j = p_j \text{ for } j \in S\}$ (see Appendix III). (The indicator function for a set takes the value 1 when the variable is in the set, and the value 0 otherwise.) Taking $f = \chi_p$ we note that the sum $\frac{1}{N} \sum_{t \in B_N} T^t \chi_p(x)$ in (1.7) coincides with the fraction

$$\frac{\text{the number of occurrences of } p \text{ in a ball of volume } N}{N} \tag{1.9}$$

in the definition (equation (1.4)) of the frequency $\nu(p)$ of the word p in x. The theorem suggests thinking of such frequencies in the form $\mathbb{I}(\mathcal{X}_p)$, that is, in terms of the invariant integral $\mathbb{I}$ and the collection C_p of tilings, a connection which we will exploit. More explicitly, even if we are primarily interested in one particular tiling $\tilde{x}$ we will find it useful to embed it in a family X of tilings, for instance in (some compact set containing) its "orbit" $O(\tilde{x}) \equiv \{T^t\tilde{x} : t \in \mathbb{Z}^d\}$, in order to use the above machinery to replace the frequency notion with which we started. In the next two sections we will consider two ways of creating such an X.

§3. Substitution tilings

As in §2, in order to study the relationship between a large number of small components within a large structure we use the common mathematical conceit that the small components are letters drawn from some finite alphabet. There is no conventional mathematical term for the large structure, which we term either a configuration or tiling, but intermediate-size structures are conventionally called "words".

In this section we develop techniques to analyze tilings made by a substitution rule such as was used for the Morse tiling of Fig. 12. This type of construction, and the analytic tools we develop here, will be used throughout the book.

So we fix some alphabet $\mathcal{A}$ and dimension d. Let $\mathcal{W}$ be the set of all possible (not necessarily finite) words made of letters from $\mathcal{A}$, namely $\bigcup_{K \subset \mathbb{Z}^d} \mathcal{A}^K$. We will call p a "subword" of the word $q \in \mathcal{A}^K$ if there is some subset $L \subset K$ such that p is the restriction of q to L. (Recall that one way to think of $\mathcal{A}^K$ is as functions from K to $\mathcal{A}$, which explains our use of the term "restriction". In this notation, the word p is a subword of the word q if at every site of $\mathbb{Z}^d$ where p has a letter q has one, and it is the same letter.) We now consider specific classes of compact subsets X of $\mathcal{A}^{\mathbb{Z}^d}$ which are "(translation) invariant", that is, such that $x \in X$ implies $T^t x \in X$ for all $t \in \mathbb{Z}^d$. Being invariant means precisely that such X carry their own representation of the translation group $\mathbb{Z}^d$, so there is no need to refer to any points of $\mathcal{A}^{\mathbb{Z}^d}$ which are not in X. An invariant and closed subset of $\mathcal{A}^{\mathbb{Z}^d}$ will be called a "subshift", as is standard in ergodic theory.

The first kind of X we consider are the "substitution subshifts". To define them we assume given a "substitution function", $F : \mathcal{A} \longrightarrow \mathcal{W}$ of the form $F = E_\gamma \circ \tilde{F}$, where $\circ$ denotes composition and $\tilde{F}$ is a map that has as range not words made of letters of the alphabet, but words made of letters which are all similar to but *smaller than* the letters in the alphabet: all shrunk by some common factor $\gamma > 1$. And E_γ is just expansion about the origin by γ.

For example, for $d = 2$ and an alphabet of four unit squares labeled A, B, C, D, we define the main component $\tilde{F}$ of the "Morse" substitution function by Fig. 10, and the similarity E_γ as in Fig. 11. This F has a natural extension to a function from $\mathcal{W}$ to $\mathcal{W}$; for instance we have, schematically:

$$F^2(A) = F[F(A)] = \begin{matrix} A & C & C & A \\ B & D & D & B \\ B & D & D & B \\ A & C & C & A \end{matrix}. \qquad (1.10)$$

Words of the form $F^k(A)$, $A \in \mathcal{A}$, will be called "letters of level k", and the set of all such words, at all levels, will be denoted $\mathcal{W}_F$. The original letters will be said to be of level 0. (It is worth noting: the four different letters of level k differ only by a permutation of the $(k-1)$-level letters in their quadrants.) We are finally ready to define the subshift X_F associated with F.

Definition 1.3. X_F is the set of all $x \in \mathcal{A}^{\mathbb{Z}^d}$ such that every finite subword of x is "special" in the sense that it is a translate of a subword of one of the words in $\mathcal{W}_F$, that is, the letters of some level.

(To preserve the relation with more general tiling theory, we will sometimes define X_F more inclusively as the set of tilings *congruent* to such elements in $\mathcal{A}^{\mathbb{Z}^d}$. The metric for that set is a bit more complicated and will be discussed in Chapter 4. It should always be clear from the context which convention we are using for X_F.)

As an example of Definition 1.3 let's take $d = 2$, $\mathcal{A} = \{A, B\}$, and

$$F(A) = \begin{matrix} A & B & A \\ B & A & B \\ A & B & A \end{matrix} \qquad (1.11)$$

and

$$F(B) = \begin{matrix} B & A & B \\ A & B & A \\ B & A & B \end{matrix}. \qquad (1.12)$$

We leave it as a useful simple exercise to compute $F^k(A)$ for a few k, from which it then follows easily that X_F consists of only two tilings, each looking like the usual red-black checkerboard (with A denoting the red squares and B the black ones), each tiling shifted one unit with respect to the other.

Any such X_F is automatically translation invariant since by the definition if $x \in X_F$, meaning that each finite subword of x is a translate of a subword of some word in $\mathcal{W}_F$, then any translate y of x also has this property and is therefore in X_F. The fact that X_F is compact is almost as simple; just recall that a closed subset of a compact space (such as $\mathcal{A}^{\mathbb{Z}^d}$) is compact, and X_F is closed since any limit $x \in \mathcal{A}^{\mathbb{Z}^d}$ of a sequence of tilings $x_n \in X_F$ can only contain words in $\mathcal{W}_F$. (This is a useful exercise.)

We now prove some simple facts about substitution subshifts, the first showing that each tiling made in this way has some form of hierarchical structure.

Lemma 1.4. *Each $x \in X_F$ is also a tiling by letters of level 1, and therefore also by letters of level k, for any fixed k.*

Proof. Fix $x \in X_F$, and consider its subwords $p^n = \{x_m : |m| \leq n\}$. ($\{m \in \mathbb{Z}^d : |m| \leq n\}$ is just the subset of $\mathbb{Z}^d$ inside a closed ball of radius n.) Recall that by construction letters of level k are made up of letters of level $k - 1$, and therefore also of all lower levels, and that from the definition of X_F we know that p^n must be a subword of a letter of some level. Therefore, except near its edge, p^n itself consists of letters of level 1. It follows that for each x_k, $|k| \leq n$, except those for which $|k| \approx n$, there is a letter of level 1 in x in which it sits; call it a_k^n. (As we vary n it needn't be true that $a_k^n = a_k^{n'}$.) So for each $M > 0$ there is an infinite subsequence n_j such that for each k, $|k| \leq M$, $a_k^{n_j}$ is independent of j. By diagonalization there is a subsequence $\tilde{n}_j$ such that for all $k \in \mathbb{Z}^d$, x_k sits in the same letter $a_k^{\tilde{n}_j} \in x$ for all j. This proves the case of level 1, and the general case follows by induction. $\square$

Visually, this says one can think of a substitution tiling as a tiling by letters *at any fixed level* – in particular, letters of arbitrarily large size. An example is exhibited in Fig. 14, where a Morse tiling is shown at two different levels. A less interesting example is shown in Fig. 15, where a periodic checkerboard tiling is shown at two levels. The latter example is less interesting because the higher-level tiling is not uniquely determined by the lower-level tiling: one can group the lower-level letters into appropriate sets of four in four different ways, leading to four different higher-level tilings. We will see from Lemma 1.9 why it is more interesting when the higher-level tiling is determined by the lower-level one, but intuitively it is reasonable that only if the correspondence between different levels in a tiling is unique does this provide a firm structural feature of the tiling which may have interesting consequences.

The next lemma gives a simple criterion for applying Theorem 1.2 to tilings. Recall that this theorem is one of the main tools for our analysis of tilings, as it enables us to understand a tiling through the frequencies of its finite subwords. To apply Theorem 1.2 we need

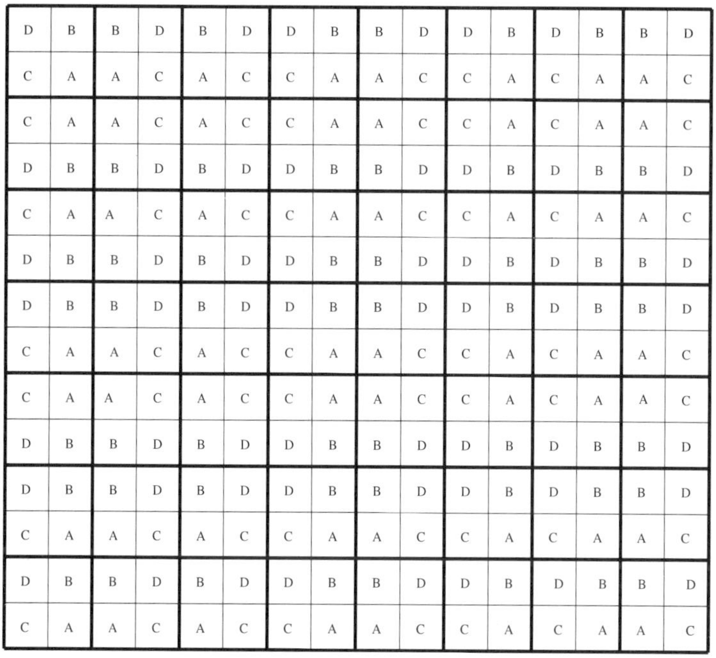

D	B	B	D	B	D	D	B	B	D	D	B	D	B	B	D
C	A	A	C	A	C	C	A	A	C	C	A	C	A	A	C
C	A	A	C	A	C	C	A	A	C	C	A	C	A	A	C
D	B	B	D	B	D	D	B	B	D	D	B	D	B	B	D
C	A	A	C	A	C	C	A	A	C	C	A	C	A	A	C
D	B	B	D	B	D	D	B	B	D	D	B	D	B	B	D
D	B	B	D	B	D	D	B	B	D	D	B	D	B	B	D
C	A	A	C	A	C	C	A	A	C	C	A	C	A	A	C
C	A	A	C	A	C	C	A	A	C	C	A	C	A	A	C
D	B	B	D	B	D	D	B	B	D	D	B	D	B	B	D
D	B	B	D	B	D	D	B	B	D	D	B	D	B	B	D
C	A	A	C	A	C	C	A	A	C	C	A	C	A	A	C
D	B	B	D	B	D	D	B	B	D	D	B	D	B	B	D
C	A	A	C	A	C	C	A	A	C	C	A	C	A	A	C

Figure 14. Two levels of a Morse tiling

to rule out a certain form of degeneracy, illustrated by the following example.

We take $d = 2$, $\mathcal{A} = \{A, \; B\}$, and define F by:

$$F(A) = \begin{matrix} A & A & A \\ A & A & A \\ A & A & A \end{matrix} \qquad (1.13)$$

and

$$F(B) = \begin{matrix} B & B & B \\ B & B & B \\ B & B & B \end{matrix}. \qquad (1.14)$$

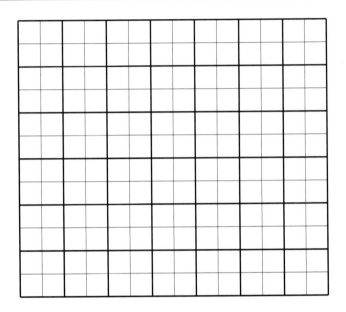

Figure 15. Two levels of a periodic tiling

$\mathcal{W}_F$ contains only words which consist only of A's or only of B's, so that each $x \in X_F$ has the same feature. That is, X_F consists of only two tilings, x_A for which all coordinates are A's and x_B with all B's, both being invariant under translation. We can then define two translation invariant integrals on X_F: $\mathbb{I}_A$ defined by $\mathbb{I}_A(f) = f(x_A)$ for all f and $\mathbb{I}_B$ defined by $\mathbb{I}_B(f) = f(x_B)$. So for this example there are two different invariant integrals and we cannot use Theorem 1.2.

The following will provide us with a simple criterion enabling us to use Theorem 1.2 in our analysis of tilings. (Note that the hypothesis does not hold for the above example.)

Lemma 1.5. *Assume there exists an integer $m \geq 1$ such that for each letter $a \in \mathcal{A}$, $F^m(a)$ contains all letters $a' \in \mathcal{A}$. Then there is one and only one translation invariant integral on X_F.*

Proof. The method of proof is to show that there is an invariant integral $\mathbb{I}$ on X_F such that condition ii) of Theorem 1.2 holds. As presented in Appendix III, we think of integrals on a compact metric space S such as X_F as certain types of linear functions on the space $C(S)$ of complex continuous functions on S. Now if we can show, for each $f \in C(X_F)$ and some tiling x', that the sequence $\frac{1}{N} \sum_{t \in B_N} T^t f(x')$ is convergent, then the limits can be used to define an integral $\mathbb{I}$ by $\mathbb{I}(f) \equiv \lim_{N \to \infty} \frac{1}{N} \sum_{t \in B_N} T^t f(x')$. Such an $\mathbb{I}$ is then easily seen to be invariant. So the only difficulty is to prove that such sequences are convergent.

Now it is a standard fact (following for instance from [Tay; p. 172]) that linear combinations of functions of the form χ_a (the indicator function for the cylinder set defined by the word a) are dense in $C(X_F)$, so we need only prove convergence for such functions. As noted in (1.9), a limit of $\frac{1}{N} \sum_{t \in B_N} T^t \chi_a(x)$ is interpretable as the frequency of the word a in the tiling x, so we are really showing the existence of such frequencies. We make one simplification: instead of dealing with arbitrary words a, we will consider only words of one letter. That is, we will show that each letter of $\mathcal{A}$ appears with a well-defined frequency, in fact the *same* frequency, in *every* tiling x. (Our proof will even show that the convergence of the approximate frequency is uniform in x, as claimed in part iii) of the theorem. The generalization to arbitrary words is not difficult, and we refer to [Ra3] for details.)

It is useful to express the alphabet in the form $\mathcal{A} = \{a_1, \ldots, a_K\}$ so we can introduce the notation that the mth iterate $F^m(a_k)$ of the substitution function produces letters (of level $m \geq 0$) of "type k". From Theorem 1.2, all we need show is that the relative fraction of letters of different type in a word of the form $F^n(a_k)$ has a well-defined limit as $n \to \infty$, and that the limit is independent of the type k. We prove this next.

We define the $K \times K$ matrix M for which M_{jk} is the number of type j letters (level 0) in a type k letter of level 1. It follows that

$(M^n)_{jk}$ is the number of type j letters in a type k letter of level n; check this! For the Morse tilings (see Fig. 10)

$$
M = \begin{pmatrix} 1 & 1 & 1 & 1 \\ 1 & 1 & 1 & 1 \\ 1 & 1 & 1 & 1 \\ 1 & 1 & 1 & 1 \end{pmatrix}. \tag{1.15}
$$

From the hypothesis there is some m such that $(M^m)_{jk} > 0$ for all j, k. (The Morse example satisfies this condition with $m = 1$.)

We now apply the Perron-Frobenius theorem, in a slightly unusual form [Rue; p. 136] (see Appendix III for notation):

Theorem 1.6. *Let $\tilde{M}$ be a real $K \times K$ matrix such that:*

    a) $\tilde{M}_{jk} \geq 0$,   for all j, k, and

    b) $(\tilde{M}^N)_{jk} > 0$,   for all j, k, for some $N \geq 1$.

Then there is a simple eigenvalue $\lambda > 0$ for both $\tilde{M}$ and its adjoint $\tilde{M}^*$, with corresponding eigenvectors $\xi, \tilde{\xi}$, having strictly positive components, and satisfying

$$
\frac{\tilde{M}^n}{\lambda^n} x \xrightarrow[n \to \infty]{} \langle \tilde{\xi}, x \rangle \xi \quad \text{for all } x \in \mathbb{C}^K. \tag{1.16}
$$

Corollary 1.7. *The spectral radius of $\tilde{M}$ (see Appendix III) is λ.*

Corollary 1.8.

$$
\frac{(\tilde{M}^n)_{jk}}{\lambda^n} = \frac{\langle \hat{e}_j, \tilde{M}^n \hat{e}_k \rangle}{\lambda^n} \xrightarrow[n]{} \tilde{\xi}_k \xi_j > 0, \tag{1.17}
$$

where $\{\hat{e}_j\}$ is the usual basis in $\mathbb{C}^K$.

We now continue with the proof of Lemma 1.5. It follows from (1.17) that there is some $\lambda > 0$, and functions g, h on $\{1, \cdots, K\}$, such that for all j, k:

$$\frac{(M^n)_{jk}}{\lambda^n} \xrightarrow[n\to\infty]{} g(j)h(k) > 0. \qquad (1.18)$$

Therefore

$$\frac{(M^n)_{jk}}{(M^n)_{j'k}} \xrightarrow[n\to\infty]{} \frac{g(j)}{g(j')} > 0. \qquad (1.19)$$

The important points are that the limits in (1.19) exist, and that they are independent of k, which are precisely what was needed to prove our result.   □

(Note that if, as in the comment after Definition 1.3, we think of X_F as all tilings *congruent* to those associated with points of $\mathcal{A}^{\mathbb{Z}^d}$ instead of just integral translates, it could not be "uniquely ergodic" – that is, have only one invariant integral – since the frequencies of tilings turned by, say, 45 degrees correspond to a different integral.)

In the next lemma we see that substitution tilings tend to be complicated, at least in the sense that they cannot have any nontrivial translational symmetry, and in particular they cannot have a unit cell such as Fig. 4. This hints at why substitutions will be so useful to us.

Lemma 1.9. *Assume the decomposition of Lemma 1.4 into letters of level k is unique for each $x \in X_F$ and level k. Then for all $x \in X_F$, $T^t x = x$ implies $t = 0$.*

We will postpone the proof of this lemma until Chapter 4 (Lemma 4.1), where we expand our analysis of symmetry properties of tilings and can then give a more general proof of this phenomenon.

§4. Finite type tilings

We saw in the last section how to use a "substitution" method to produce rather complicated tilings. But that method is not a *local* rule of the type we want. We are trying to understand what kind of

Figure 16. The kite and dart

global structures, for instance tilings, can be produced by local rules like those of a jigsaw puzzle, rules which only depend on parts of the global structure of limited size. The rules of a jigsaw puzzle consist merely of the requirement that neighboring tiles fit together, which is local in this sense, while the tilings made by substitution require consideration of subwords of all sizes, and this is nonlocal.

Let us go back to a primary example, the kite & dart tilings of Fig. 1. These are composed of two simple shapes, a kite and a dart (Fig. 16). (The kite is named after the toy, and the dart after the sewing term.) There is a substitution rule for the kite & dart tilings, using $\tilde{F}$ in Fig. 17 with $\gamma = (3 + \sqrt{5})/2$. (The dots form the outlines of the kite and dart, which are associated by $\tilde{F}$ with several small size darts and kites, outlined in solid lines in the figure.) It is not hard to show that the substitution function $F = E_\gamma \circ \tilde{F}$ can be extended to map words to words (as is automatically true for subshifts); that is, the image of a word is again a word, without any overlapping. The kite & dart (substitution) tilings are now defined roughly as in the previous section, as all tilings x such that each finite word in x is congruent to a subword of some $F^m(a)$ where a is a kite or dart.

But there is a "better" way to understand kite & dart tilings, using the two tiles in Fig. 18, and jigsaw puzzle rules. That is, it can be proven [Gar] that the *only* way to tile the plane with copies of the tiles in Fig. 18 is to make a tiling in which these modified tiles *must* be grouped together into the collections indicated in Fig. 17;

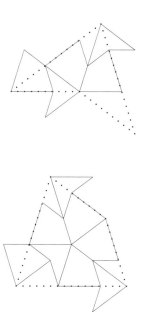

Figure 17. Substitution $\tilde{F}$ for kite and dart

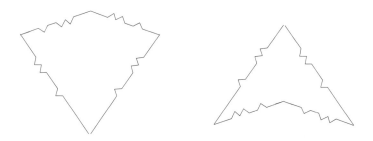

Figure 18. The modified kite and dart

and these collections *must* be grouped together into the collections of collections produced by F^2, etc. In other words, one can modify the original quadrilaterals of Fig. 16, adding bumps and dents to the edges as in Fig. 18, so that the *only* way these modified tiles can tile the plane is in the special ways the unmodified tiles tiled the plane using the substitution method. The two original quadrilaterals of Fig. 16 can abut to form a rhombus from which one can tile the plane in a simple periodic way; this is what adding the bumps manages to avoid.

This is magic! Remember, the substitution method was a simple way to produce complicated tilings. And what we have here is a way to reproduce the interesting end result, but now using honest-to-goodness local rules! And not only is there such a trick for the kite & dart tilings but for "most" substitutions. This is important for our analysis so we will next go through a more complicated example in detail, the Morse tilings, which we produced earlier by a substitution process. That is, we will now show how to make new tiles, which are unit squares like the originals but with certain patterns of bumps and dents on their edges, so that they can *only* tile the plane in the manner of the Morse substitution tilings.

The original Morse set (Fig. 9) contains four tiles and the new set will contain 56, so the situation is not as simple as for the kites and darts, where there was only one new tile for each original tile. For the kite & dart tilings it is easy to see why we say the modified tiles reproduce the original tilings – all you have to do is ignore the bumps and dents of the modified tiles in one of their tilings and you get an original tiling. The situation has to be a bit more complicated with the Morse tilings, and the best way to understand this is through the proof that will follow.

Before describing the new bumpy tiles, let's review some qualitative features of the Morse tilings. There are four different Morse tiles, which we can think of as differently colored unit squares, called A, B, C and D. From Lemma 1.4 the Morse tilings can be thought of as consisting of nonoverlapping *collections* of letters or tiles, each collection being one of the four shown in Fig. 10. In other words, a

Morse tiling can be interpreted as in Fig. 14, a tiling by letters of level 1. Continuing this line of thought, these letters of level 1 fit together to make letters of level 2, which are collections of 16 ordinary tiles (as in equation (1.10)), leading to a different interpretation of the same Morse tiling, as a tiling by letters of level 2: and so on. It will be useful to know that these decompositions, of a Morse tiling into letters of arbitrary but fixed level, are unique in the sense of Lemma 1.9. To see this, note (as one sees in Fig. 12) that each row and each column in a Morse tiling consists of a sequence of only two of the letters. Since in each of the letters of level 1 no letter is repeated, it follows that in a row or column of a tiling wherever one finds a letter repeated the pair cannot belong to the same letter of level 1. So once we find such a repetition in a row we know where the vertical boundaries of one, and therefore all, letters of level 1 must be, throughout the tiling. And once we find such a repetition in a column we know where the horizontal boundaries of all letters of level 1 must be. (All rows and columns have repetitions.) So there is only one way to decompose a tiling into letters of level 1. And similarly there is only one way to decompose a tiling into letters of level 2, level 3, etc.

In fact one can think of these tilings at different levels in a Morse tiling as having been constructed by a sequence of choices as follows. "Choose" some starting letter, say A, at the origin of $\mathbb{Z}^2$. Then choose one of the four possible letters of level 1 for it to belong to, say that associated with letter B – what we call the letter of level 1 and type B. The letter A is in the bottom left corner of that collection, which tells us where three other (level 0) letters are to be in the tiling. Now think of how this letter of level 1 and type B must sit in a letter of level 2 – that is, choose another of the four types, say A, again. That tells us that the level 1 letter sits in the upper left corner of the collection of 16 letters, which determines 12 more (level 0) letters of the tiling. In this way, by a sequence of choices we determine the tiling.

We should point out however that there is one complication in the above analysis; not every sequence of choices leads to a full tiling. For instance, if we choose type A every time, we only fill out a quadrant, not the whole plane. So some tilings require two, three or four sequences of choices to fill out the plane. Given that complication, this is a way to classify all the Morse tilings, and is a useful way to understand the Morse tilings, or indeed any substitution tiling (at least when Lemma 1.9 applies).

We have reviewed the Morse tilings sufficiently, and now we return to our promise to exhibit a new set of tiles (due to Raphael Robinson ([Rob, GrS])) which can *only* tile the plane in the manner of the Morse tilings. That is, we will construct a finite alphabet $\mathcal{A}$ of square-like tiles, and consider the set $X_{\mathcal{A}}$ of *all* tilings that can be made by congruent copies of the letters in $\mathcal{A}$, and show that in an appropriate sense these are the "same" as the Morse tilings. Systems such as $X_{\mathcal{A}}$ will be called "finite type subshifts", to distinguish them from the substitution subshifts of the last section. We record some of this in the following definition.

Definition 1.9. Given a finite alphabet $\mathcal{A}$ of polyhedra in $\mathbb{R}^d$, we define the "finite type" system $X_{\mathcal{A}}$ as the set of *all* tilings of $\mathbb{R}^d$ by congruent copies of letters in $\mathcal{A}$.

We are going to think of $\mathcal{A}$ as a rule for producing tilings. This may seem odd at first, since by generating "all" tilings made from an alphabet one expects the output to contain a wide variety of tilings, inappropriate for the intuitive expectation for a production rule. But this need not be the case. The first tilings we constructed in the Introduction, using an alphabet of K^2 tiles, only allows one tiling up to congruence (pictured in Fig. 4 for $K = 6$). And we will be concentrating on alphabets which only allow tilings which are locally indistinguishable – every finite subword of one appears in every tiling made from that alphabet – so in an intuitive sense one can still think of the output as being essentially unique, that is, as the rule "producing" some essentially unique thing. This is also the manner in which the laws of statistical mechanics picture the production of bulk matter, as we will see in the next chapter.

If the alphabet is square-like then as with substitution subshifts we can restrict attention to those tilings for which the centers of the tiles lie on $\mathbb{Z}^d$. Then a finite type subshift $X_{\mathcal{A}}$ is a translation invariant, compact subset of $\mathcal{A}^{\mathbb{Z}^d}$ for the same reasons as for substitution subshifts.

Now for Robinson's example. (This argument is a bit long and is not used in the remainder of the text, so it could be skipped on first reading.) Consider first the five "basic" tiles at the top of Fig. 19. Number 1 is called a "cross", and the other four are called "arms". For later interpretation it will be convenient to substitute the tiles directly below them, which replace the bumps and dents by appropriate arrows.

Consider also the three "parity" tiles with bumps and dents, numbered 6-8, and the three tiles below them which substitute arrows for the bumps and dents.

The tiles we will actually use are obtained by superposing these eight preliminary tiles (in the arrow versions). With tile 1 one can superimpose tile 7. With tiles 2–5 one can superimpose tile 8. And any of tiles 1–5 can be superimposed with tile 6. Those are the rules for superimposing, and they yield ten different tiles as in Fig. 20. In producing tilings, each of these ten types of tile can be rotated or reflected. (A little thought shows these lead to 56 different types of tile which must be used if one is not allowed to rotate or reflect in producing a tiling).

Remember that the arrows stand for bumps and dents, so it is easy to determine which pairs of tiles may abut in a tiling: only those for which arrow heads meet arrow tails and vice versa.

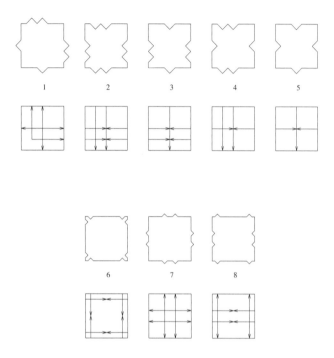

Figure 19. The five basic tiles and three parity tiles

The patterns of arrows have been separated into two classes for a reason. Each tile consists of a basic component and a parity component, and the rule which determines which pairs of tiles may abut in a tiling reduces to requiring that the pair of basic components may abut together with requiring that the pair of parity components may abut.

The parity tiles are simpler so we consider them first. In a tiling parity components must alternate horizontally and vertically in the pattern shown in Fig. 21. So much for the parity components!

To analyze how the basic components of tiles may appear in a tiling we consider two aspects. First we use the rules by which parity

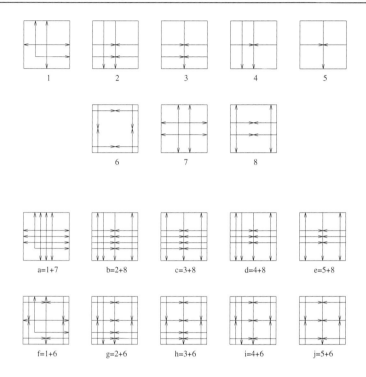

Figure 20. Constructing the ten Robinson tiles, $a - j$

Figure 21. Unit cell of parity tiles

and basic components are combined to see what follows from the alternating pattern just determined for parity components. Now from the fact that parity tile 7 only combines with a cross, and from the alternating pattern of parity components, we see that in a tiling certain crosses (those combined with parity tile 7, not all crosses) will appear in every other position horizontally and every other position vertically. These tiles will be called 1-squares. The remainder of the analysis of the basic components will not depend on any other feature of the parity components.

For convenience we introduce some notation and abbreviated graphics for the basic tiles. Arrows on these tiles will be called "in arrows" if they end in the interior of the tile, and "out arrows" if they end at an edge. An arrow will be called "central" if it lies on a line going through the center of the tile, and otherwise it is a "side arrow". Each of the four types of arm has one "principal" arrow, the central out arrow. We will sometimes abbreviate an arm by displaying only its principal arrow or its principal and side out arrows, as in the top of Fig. 22.

The cross in Fig. 19 is said to "face" in two directions, to the right and upward, because of the directions of its side arrows. When we wish to ignore this feature we abbreviate the cross as in the bottom of Fig. 22.

We now begin the analysis of the pattern of basic components of tiles with the 1-square. Straightforward checking of possibilities shows that they must appear at the corners of sets of nine tiles as in Fig. 23. (Although we know there is a cross at the center, we do not know in which directions it faces.) Such a set of nine tiles with 1-squares at the corners is called a 3-square, and it will be said to face in the directions of the cross at its center.

We next determine the pattern of those crosses which are not 1-squares. (If we use the suggestive notation that the 1-squares are in the "odd-odd" positions in the tiling, then these other crosses can only be in certain "even-even" positions.) Suppose we have such a cross in

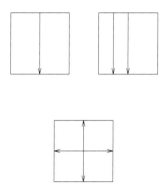

Figure 22. Abbreviated basic tiles

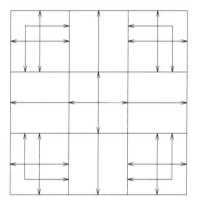

Figure 23. A 3-square

some row, with one side arrow facing up, the other facing either left or right. Consider the tiles to the right of the cross. There will first be some sequence (possibly empty) of horizontal arms headed right. If the sequence is not infinite, there must then be a single vertical arm, then a sequence (possibly empty) of horizontal arms headed left, and another cross. Because of their side arrows, two such consecutive

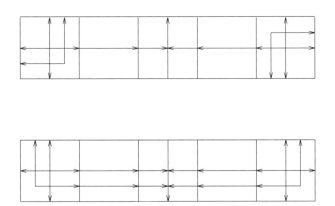

Figure 24. Consecutive crosses

crosses must either be back-to-back as in the top row of Fig. 24, or face each other as in the bottom row of Fig. 24.

We will next show that for consecutive crosses which face one another the separation is even, measured between the centers of the tiles. Consider the horizontal arms and single vertical arm between such a pair of crosses. They all have arrow tails on their top edges, so the tiles abutting their top edges must consist alternately of crosses and vertical arms headed down. Since this argument also holds for columns, this alternating sequence of crosses and vertical arms must begin and end with crosses, as in Fig. 25. This proves that the separation between facing crosses is even.

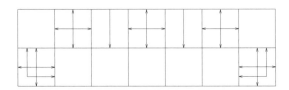

Figure 25. An even distance between crosses

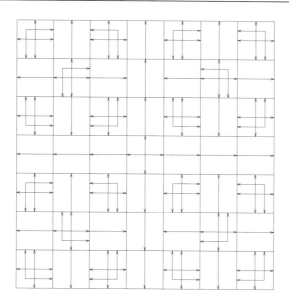

Figure 26. A 7-square

Let us now apply these facts about facing crosses to the crosses at the centers of 3-squares. Suppose such a cross faces up and to the right, as in the bottom left corner of Fig. 26. We will show that it determines the other tiles of Fig. 26 to the extent indicated. Note first that the directions of the three arms to the left and the three arms below the tile in the center of the figure are forced by the side arrows of the cross in the center of our 3-square. These arms then force a cross to be in the center of the figure. From our previous analysis this cross fixes the positions (but not directions) of the crosses in the row above it and in the column to its right. The directions of these crosses however must be as shown, for otherwise there would be the head of an arm abutting the central cross. And these crosses in the neighboring row and column force the three 3-squares as well as the rest of the arms. In this way each of the four 3-squares in Fig. 26 forces the remaining pattern, which we call a 7-square.

Similarly, each 7-square forces the presence of three other 7-squares and the rest of a 15-square by expanding in the directions of the central cross of the first 7-square. And similarly for 31-squares, 63-squares, etc. In summary, each 1-square in a tiling is contained within a unique 3-square, which is contained in a unique 7-square, and so on for $(2^n - 1)$-squares of all n. This expanding sequence of squares can be thought of as being produced by a sequence of choices of directions for the central crosses.

However, starting from a 1-square and making choices may lead to a quarter plane (for instance, if all choices of pairs of directions are the same), or a half plane (for instance if half the choices are up and to the right, and the other half are up and to the left). So a tiling may be made up of four such quarter planes, or two such half planes, or a half plane and two quarter planes. Between two such half planes or quarter planes there must be a row or column of arms. Now although 1-squares appear in a perfect checkerboard pattern because of the parity tiles, this need not be the case for the other $(2^n - 1)$-squares. Specifically, this may break down across the column or row dividing the quarter or half planes we have been discussing; consider the 3-squares in Fig. 27. Such a column or row will be called a "fault line".

Note that except for tilings with fault lines, the tilings we have described for these ten Robinson tiles have a natural one-to-one correspondence with the Morse tilings. That is, we recall that the Morse tilings can be thought of as being created by an infinite sequence of choices; as one builds a larger and larger part of the tiling, each term in the sequence of choices determines on which of the four corners of a new larger square the already produced collection of tiles will lie. And of course the Robinson tilings can also be built that way, with the choice of directions for the nth cross determining which corner of the $(2^n - 1)$-square the already built $(2^{n-1} - 1)$-square will occupy. We will have to discuss the nature of this correspondence between the two sets of tilings in more depth, to determine which features it

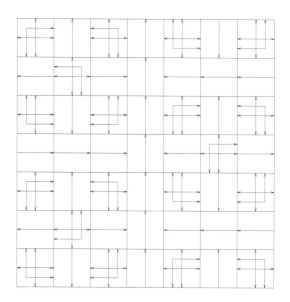

Figure 27. The central column is a fault line

preserves and which it does not, and the relevance of these various features. (Is it a problem that the number of types of tiles is different in the two collections of tilings?)

Recalling some notation from the Introduction, if we let $\mathcal{A}$ be the set of ten Robinson tiles $a - j$ in Fig. 20 and $X_{\mathcal{A}}$ be the set of all tilings that can be made with these tiles, we have shown that, except for some complication about tilings which decompose into tilings of quarter and/or half planes, there is a natural one-to-one correspondence between the Robinson tilings $X_{\mathcal{A}}$ and the Morse tilings X_F. We have introduced this notation about sets X of tilings, with a metric on X and integration of functions on X, just for such analysis, to which we now proceed.

Consider first the Morse substitution system. It satisfies the conditions for Lemma 1.5 since every letter of level 1 contains all four letters of $\mathcal{A}$, so we know there is only one $\mathbb{Z}^2$-invariant probability integral on X_F. For the Morse X_F (resp. Robinson $X_{\mathcal{A}}$) consider the "nice" subset $\tilde{X}_F$ (resp. $\tilde{X}_{\mathcal{A}}$) consisting of those tilings associated with only one sequence of choices (i.e., for which the sequence determines a tiling of the full plane, not just a quadrant or half space).

Recall the metric we have placed on spaces of tilings. Two tilings are close in this metric if and only if as functions on $\mathbb{Z}^2$ they agree on all points in $\mathbb{Z}^2$ in a large ball centered at the origin. Recalling the correspondence between $\tilde{X}_F$ and $\tilde{X}_{\mathcal{A}}$ it follows that the one-to-one correspondence, and its inverse, are continuous. We will show that in a natural sense the rest of the spaces X_F and $X_{\mathcal{A}}$ are negligible, but first a few words about the negligible sets.

Consider the frequency with which some finite word w appears in a tiling x in $\tilde{X}_F$ or $\tilde{X}_{\mathcal{A}}$. Making use of the hierarchical structure of x – the decomposition into letters of any fixed level for $\tilde{X}_F$, or into $(2^n - 1)$-squares for $\tilde{X}_{\mathcal{A}}$ – we can, as in the proof of Lemma 1.5, estimate the frequency of some fixed w by computing its frequency in a letter of high level if $x \in \tilde{X}_F$ or its frequency in some $(2^n - 1)$-square for large n if $x \in \tilde{X}_{\mathcal{A}}$. And as shown in that proof, such a frequency is well-defined in the limit of large hierarchical level, and independent of the type of letter (or $(2^n - 1)$-square) we use. Precisely the same words appear in all the $x \in \tilde{X}_F$ or in all the $x \in \tilde{X}_{\mathcal{A}}$. This may not be true for x outside these "nice" sets. For instance if $x \in X_F - \tilde{X}_F$ contains a fault line as defined above, there are words which straddle the fault line and which do not appear in any $x \in \tilde{X}_F$ (Fig. 27). However such a word can only appear along a fault line, so its frequency is 0.

Now fault lines are by definition boundaries between "infinite-level" parts of the hierarchy; they are boundaries at which two infinite parts do not meet well, in the sense that there are words overlapping the boundaries which are not subwords of letters of some level. For completeness we note that we can have infinite parts meeting well, as occurs naturally in substitution systems. Recall that a substitution system X_F consists of all tilings which *only* contain words which are subwords of letters of some level, so there cannot be such a thing as a fault line in a substitution system. But there will be tilings in X_F which contain parts of "infinite-level" letters, as follows. Take any tiling x in the nice part $\tilde{X}_F$ of a substitution system. It contains letters L_n of level n and type α for all n and α. Construct the new tiling x_n by translating x so that a letter of level n and some specific type $\tilde{\alpha}$ has the tile of its bottom left corner at the origin. Although this sequence $\{x_n\}$ of tilings may not have a limit, any accumulation point x in X_F of $\{x_n\}$ will have part of an infinite-level letter in its first quadrant. (In general the *type* of that infinite-level letter is not meaningful since two letters of a given level only differ on the largest scale; so there can only be one type of infinite-level letter.) The existence of such accumulation points x follows from the fact, noted just after their definition, that substitution systems such as X_F are compact.

Finally, we said we would show that the "nice" tilings are the essence of the examples, that the rest of the tilings are, in some reasonable sense, negligible. We show this simultaneously for the Morse and Robinson systems; we call the full system of tilings X, the nice subset $\tilde{X}$ and the invariant integral $\mathbb{I}$.

Our definition of a "negligible" set S is one for which for any $\epsilon > 0$ there is a set $S_\epsilon \supseteq S$ such that $\mathbb{I}(\chi_{S_\epsilon}) < \epsilon$. Recall that the indicator function of a set, such as $\chi_{S_\epsilon}(x)$, has value 1 when the variable $x \in S_\epsilon$ and value 0 otherwise. The integral of such a function is suggestive of the size of the set. (In particular, the integral of such a function must decrease with the set – if $S' \subset S''$ then $\chi_{S'} \leq \chi_{S''}$ and so $\mathbb{I}(\chi_{S'}) \leq \mathbb{I}(\chi_{S''})$ – and if for some family $\{S(n)\}$ we have $\bigcap_n S(n) = \emptyset$, then, with $S^N \equiv \bigcap_{n \leq N} S(n)$, $\mathbb{I}(S^N) \to 0$ as $N \to \infty$ by the Monotone Convergence Theorem – Appendix III.) So the condition $\mathbb{I}(\chi_{S_\epsilon}) < \epsilon$ indicates that the set S_ϵ is small, and we are saying that a set is negligible if there are such arbitrarily small sets containing it.

Now although it is quite possible to have negligible sets containing uncountably many points, it is always true (if the full space X is uncountable) that a countable subset $S \subset X$ is automatically negligible. To see this note first that the indicator functions of any two singletons (sets containing only one point) must have the same integral since the integral is invariant. If this invariant value were not zero then the indicator function χ_{S_n} for any set S_n containing a suitably large finite number of points would, by linearity of the integral, have integral $\mathbb{I}(\chi_{S_n}) > 1 = \mathbb{I}(\chi_X)$. But this would contradict the fact that $\chi_{S_n} \leq \chi_X$. Now, given a set $S \subset X$ containing countably many points, we can choose a sequence of subsets S_n containing n points, $S_n \subset S$, $S_{n+1} \supset S_n$ and $\bigcup_n S_n = S$. Then from the Monotone Convergence Theorem (Appendix III), $0 = \mathbb{I}(\chi_{S_n}) \longrightarrow \mathbb{I}(\chi_S)$ and $\mathbb{I}(\chi_S) = 0$, as claimed.

Getting back to the tilings, we want to show that $X - \tilde{X}$ is negligible. So consider a point in $X - \tilde{X}$, for instance a tiling with an infinite vertical line, corresponding to some infinite-level letter or infinite-level square on at least one side. What can be said of the set A of all such tilings? A is the countable union of *disjoint* closed sets A_n in which the line has fixed position indexed by n. (These sets are disjoint since a tiling cannot have two such infinite edges – the region in between would be incompatible with the unique hierarchical structure of the tilings, which we analyzed early in this chapter.) For each of these countably many subsets A_n the indicator function χ_{A_n} has the same integral, since the sets are translates of each other and $\mathbb{I}$ is translation invariant. By the same method as before we see that this value must by 0, and that the integral $\mathbb{I}(\chi_A)$ of the indicator function of the union $A = \bigcup_n A_n$ must also have value 0.

This is the basic argument. Now there are a finite number of other such cases – tilings with various combinations of quarter spaces and half spaces – and by the same reasoning the set of such tilings is negligible. This completes the argument that the nice tilings are the essence of the examples, and our demonstration of how a substitution system can be mimicked by finite type (i.e., jigsaw puzzle) rules. We

Figure 28. The Dekking-Keane tiles

have intimated that this finite type property is wonderful, so we next show how it can be used.

Imagine we wanted to make a set of tiles which could only tile the plane in a disorderly manner, in the sense of section 1 of this chapter. For various reasons this sort of study has been made in some depth but not for tilings made by the finite type rules in which we are interested. For instance, it is known [DeK] that the Dekking-Keane substitution tilings made with the tiles of Fig. 28 and the substitution rules of Fig. 29 are disorderly to the extent that they satisfy (1.3). Now if we had a general way to mimic with finite type rules the tilings made by substitution rules, in the way we just mimicked the Morse tilings with Robinson tilings, we would be in business. To a large extent such a general procedure is known, due to Shahar Mozes [Moz]. We will not give his results in complete generality, but restrict ourselves to the most useful situation.

We begin with "symbolic substitutions", consisting of a finite alphabet of abstract symbols, and, associated with each letter in the alphabet, a word of at least 2 letters. For instance the alphabet might be $\mathcal{A} \equiv \{0, 1\}$ and the associated words (containing 3 and 5 letters respectively):

$$0 \longrightarrow 001, \quad 1 \longrightarrow 11100. \tag{1.20}$$

Given 2 such alphabets $\mathcal{A}_1 = \{a_1, \cdots, a_K\}$ and $\mathcal{A}_2 = \{c_1, \cdots, c_L\}$, and substitutions

$$
\begin{aligned}
a_j \in \mathcal{A}_1 &\longrightarrow a_{k_1(j)} a_{k_2(j)} \cdots a_{k_{n(j)}(j)}, \\
c_k \in \mathcal{A}_2 &\longrightarrow c_{j_1(k)} c_{j_2(k)} \cdots c_{j_{m(k)}(k)},
\end{aligned}
\tag{1.21}
$$

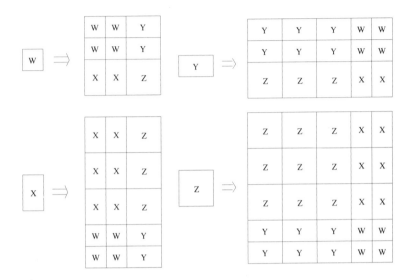

Figure 29. The Dekking-Keane substitution

we construct an alphabet of rectangles in the plane and a substitution function for that alphabet. The alphabet has one rectangle for each element of the product $\mathcal{A}_1 \times \mathcal{A}_2$; namely, associated with $(a_j, c_k) \in \mathcal{A}_1 \times \mathcal{A}_2$ we have the rectangle $R(a_j, c_k)$ with dimensions $n(j) \times m(k)$ (taken from the substitution rules (1.21). The substitution for the rectangle $R(a_j, c_k)$ is made by using the substitution for $a_j \in \mathcal{A}_1$ horizontally and the substitution for $c_k \in \mathcal{A}_2$ vertically:

$$
\begin{array}{cccc}
R(a_{k_1(j)}, c_{j_1(k)}) & R(a_{k_2(j)}, c_{j_1(k)}) & \cdots & R(a_{k_{n(j)}(j)}, c_{j_1(k)}) \\
R(a_{k_1(j)}, c_{j_2(k)}) & R(a_{k_2(j)}, c_{j_2(k)}) & \cdots & R(a_{k_{n(j)}(j)}, c_{j_2(k)}) \\
\cdots & \cdots & \cdots & \cdots \\
\cdots & \cdots & \cdots & \cdots \\
\cdots & \cdots & \cdots & \cdots \\
R(a_{k_1(j)}, c_{j_{m(k)}(k)}) & R(a_{k_2(j)}, c_{j_{m(k)}(k)}) & \cdots & R(a_{k_{n(j)}(j)}, c_{j_{m(k)}(k)})
\end{array}
$$

$$(1.22)$$

In particular, the Dekking-Keane substitution of Fig. 29 is made from 2 copies of the symbolic substitution of (1.20); check this! Given any such substitutions (1.21), Mozes gives a prescription for producing a finite type system which mimics the tilings of the rectangles (1.22). We are now in a position to state the theorem of Mozes. To reiterate, it is a generalization of what we did for the Morse substitution of Fig. 10; the theorem takes a rather general substitution system and reproduces the global structures of that system with structures built by local rules. Thus the theorem shows how to reproduce the complicated structures, which are easily obtainable by substitution rules, by the type of rules we want, local (jigsaw-like) rules.

Theorem 1.10 (Mozes). *Suppose we are given a substitution subshift X_F with unique invariant integral $\mathbb{I}_F$ made as above from two nonperiodic symbolic substitutions. There is a finite alphabet $\mathcal{A}$ and finite type subshift $X_{\mathcal{A}}$ with unique invariant integral $\mathbb{I}_{\mathcal{A}}$ with the following property. Ignoring a negligible set in X_F and one in $X_{\mathcal{A}}$, there is a bicontinuous one-to-one map ϕ between X_F and $X_{\mathcal{A}}$ which commutes with translation, i.e., $\phi[T^t(x)] = T^t[\phi(x)]$ for tilings x and translations T^t.*

(As with the Robinson and Morse tilings, the exceptional negligible sets correspond to tilings containing infinite lines.) The correspondence produced by Mozes – and we note that the method is constructive, it is not just an existence theorem – ensures, for instance, that the tilings in the two systems have precisely the same properties of order or disorder, since the map ϕ shows that the abstract features of the two systems, which determine the order properties, are isomorphic.

This has been a long chapter, so it would be useful to recapitulate a few key points. In the first half we showed how a probabilistic approach to the order properties of a tiling leads, via Theorem 1.2, to embedding the tiling of interest in a family of tilings, with a probability distribution on the family. (This discussion will be greatly

expanded in Chapter 3.) The second half of this chapter was devoted
to two kinds of tilings, those made by a substitution method and
those like a jigsaw puzzle, made by local rules. These two types of
tilings were then related by the theorem of Mozes. The Mozes machinery is quite powerful, and enables us to produce examples with
rather strong properties of disorder. We will explore this consequence
in some depth in Chapter 3, but we first want to look at our situation
from another (physics) angle, which will give us a stronger intuition
from which to understand our structures.

Chapter 2

Physics (for Mathematicians)

§1. Diffraction of light waves

One of the main threads running through this book is the goal to understand what kind of global structures can be produced from local rules, rules such as that of a jigsaw puzzle – which determine a global structure merely by requiring that neighboring tiles fit together correctly. In this chapter we will develop a different perspective on this subject. As discussed in the Introduction, physics has been led to one version of such an analysis in order to understand the global structure of solid matter made from atoms. That endeavor has been colored by the availability of X-ray diffraction as a tool in exploring such structures. In performing diffraction one takes the sample to be examined and bombards it with waves of some kind; for X-ray diffraction one uses light waves, so we will begin with a brief introduction to light waves and their diffraction.

First some notation. By a "field" in physics one means a function on $\mathbb{R}^3$; a "scalar field" has values in $\mathbb{R}$ or $\mathbb{C}$, and a "vector field" has values in $\mathbb{R}^3$ or $\mathbb{C}^3$. A "time-dependent field" is likewise a (scalar or vector-valued) function on $\mathbb{R}^3 \times \mathbb{R}$ (space $\times$ time). A "plane wave" is a special case of a time dependent field. A (complex) scalar plane wave is any function of the form:

$$f(x,t) = Ae^{i(k \cdot x - \omega t + a)}, \qquad (2.1)$$

where the constants $A \in \mathbb{R}_+$, $k \in \mathbb{R}^3$, $\omega \in \mathbb{R}_+$ and $a \in \mathbb{R}$ are called, respectively, the "amplitude", "wave vector", "frequency" and "phase" of the wave. To understand these terms, and the term plane wave, simply consider the level surfaces of f for constant time t: they are planes perpendicular to k, and they move in the direction of k as t increases. Note that physics usually uses complex plane waves with the understanding that the physically relevant quantity is the real part of the function. This goes for vector fields also.

Light waves, of which X-rays are special cases, are a combination of two time-dependent vector fields, the electric $E(x,t)$ and magnetic $B(x,t)$ fields, which satisfy Maxwell's equations [EiI]. Only $E(x,t)$ contributes to X-ray diffraction, so we will ignore $B(x,t)$ – and Maxwell's equations.

A plane wave of the electric field is a field of the form:

$$E(x,t) = e_1 A_1 e^{i(k \cdot x - \omega t + a_1)} + e_2 A_2 e^{i(k \cdot x - \omega t + a_2)}, \qquad (2.2)$$

where e_1, e_2, $k \in \mathbb{R}^3$ are pairwise perpendicular (and constant, i.e., independent of x and t), so e_1 and e_2 are in the plane perpendicular to k (which makes light waves "transverse"). Such a wave is "elliptically polarized" because the endpoint of (the real part of) the vector $E(x,t)$ traces out an ellipse in time at any fixed x in space. Special cases are "linear polarization" when $a_1 = a_2$, and "circular polarization" when $A_1 = A_2$ and $|a_1 - a_2| = \pi/2$.

In a typical diffraction experiment we arrange (roughly) to have a plane wave of given wavelength directed at a location containing some sample, and to measure the plane wave that scatters off the sample, in any given direction. Before trying to analyze this for realistic samples, we consider the simplest case wherein the sample consists of a single electric charge (electron).

So assume a (scalar) plane wave

$$E(r,t) = E_0 e^{i[2\pi\nu(t - r_0 \cdot r/c)]} \tag{2.3}$$

(where r_0 is a unit vector in the direction of motion of the wave) scatters off a charge at the origin and the scattered field at (r,t) is

$$f(r)E_0 e^{i[2\pi\nu(t - |r|/c) - \psi]} \tag{2.4}$$

The main feature of this quantity is seen from its level surfaces, which are spheres. As time advances the sphere expands, at the speed c (the speed of light). ψ just governs the relative phase between the incoming plane wave and the scattered sphere. For a free electron $\psi = \pi$, and the scattering factor $f(r)$ can be computed [Gui; p. 9]. (The derivation of f is the only place it is desirable to consider the vector nature of the field, and since we are ignoring this aspect we can work with a scalar field.)

Next assume we have an incoming plane wave as above but now assume it scatters off two electrons near our origin separated by u. At r the difference in phase between the two scattered waves is $-2\pi s \cdot u$ where $s = (r/|r| - r_0)/\lambda$ and $\lambda = c/\nu$ is the wavelength of the scattered waves. Therefore the net field at r scattered from the incoming beam by a collection of electrons with density function ρ_V in a region V has, relative to that of a free electron at the origin, an amplitude given by

$$E_0 \int_{\mathbb{R}^3} \rho_V(u) e^{-2\pi i s \cdot u} \, du, \tag{2.5}$$

and an *intensity*

$$I_V(s) = \left(E_0 \int_{\mathbb{R}^3} \rho_V(w) e^{2\pi i s \cdot w} \, dw\right)\left(E_0 \int_{\mathbb{R}^3} \rho_V(v) e^{-2\pi i s \cdot v} \, dv\right), \tag{2.6}$$

which by change of variables is

$$(E_0)^2 \int_{\mathbb{R}^3} [\int_{\mathbb{R}^3} \rho_V(u+u')\rho_V(u')\,du']e^{-2\pi i s \cdot u}\,du. \qquad (2.7)$$

This intensity I_V is the physically accessible quantity that diffraction gives us, to help us understand the atomic structure of bulk matter. We will make use of it through its Fourier transform, a more convenient quantity which goes by the name of the 'autocorrelation" A_V:

$$A_V(u) = \int_V \rho_V(u'+u)\rho_V(u')\,du' = \int_{\mathbb{R}^3} \rho_V(u'+u)\rho_V(u')\,du'. \qquad (2.8)$$

One can read off from (2.8) the following features of the autocorrelation. It samples the distribution of electrons throughout the region V, and the main point is that $A_V(u)$ has larger values for those vectors u which separate sizable populations of electrons. Also, it is an average quantity, responsive to the total population of electrons, not the details of each small region or the location of any particular electron. Since we are interested in the details of the small-scale structure of our patterns, this is a limitation. As we shall see next, this limitation of the physical measurability of the atomic structures of bulk matter also appears at the theoretical level.

§2. Statistical mechanics

We will review here the standard model for understanding the basic properties of matter under simple conditions, so called equilibrium statistical mechanics. Assume we wish to analyze the physical properties of a piece of homogeneous matter composed of many atoms (for simplicity we assume they are all of the same element, say iron), at various fixed temperatures $1/\beta > 0$ and chemical potentials c – physically adjustable parameters (which one needn't "understand" for the following). To be more specific, imagine we want to show that at some temperature and chemical potential statistical mechanics predicts that iron atoms should cluster together into a "crystal" – by which we mean a configuration of atoms (or more properly, atomic nuclei) arranged spatially as in Fig. 2. Keep this in mind as we now give an outline of statistical mechanics.

Assume there are $N \gg 1$ iron atoms contained in an otherwise empty cube V. We will think of their positions as N variable points in V, or as a point in V^N. We assume further that we know how to assign a "(potential) energy" $E^V(x; c)$ to each possible configuration x of V^N. This is usually done by adding together contributions from each pair of points in the configuration. (We are ignoring some features of our iron atoms which are relatively minor and easy to insert if desired.) Then for fixed β and c we consider

$$f^{\beta,c}(x) \equiv \frac{\exp[-\beta E^V(x; c)]}{\int_{V^N} \exp[-\beta E^V(x; c)] \, d^N x} \qquad (2.9)$$

as a probability density for x. The essence of statistical mechanics is to treat this probability density as the fundamental quantity from which one computes properties of the material. Statistical mechanics does not attempt to tell us precisely what the configuration of the atoms is; it only tells us which, among all conceivable configurations, are the more "likely" ones. Then, to compute something about the system (such as the density autocorrelation of the last section), we compute it for each conceivable configuration, and then average the results using the above probabilistic weights of the configurations. That is, for any property that is a function $g(x)$ of the configuration, the theory predicts the value $\int_{V^N} g(x) f^{\beta,c}(x) \, d^N x$. For instance, the (potential) energy of the system would be $\int_{V^N} E^V(x; c) f^{\beta,c}(x) \, d^N x$.

Returning to the motivation of the beginning of this section, imagine we want to show that at some temperature and chemical potential, and with some energy function $E^V(x; c)$, the theory predicts the iron atoms form a "crystal", a configuration of atoms arranged spatially as in Fig. 2. How could this come out of such a probabilistic theory? Well, to be painfully honest no one has actually shown, using realistic conditions, that such a prediction follows from the theory. But very simplified models suggest it would, in the following way. The theory only gives a probability distribution among all possible configurations. But it is expected that if one computes for a system containing an enormous number of atoms (and in a piece of bulk matter there are indeed roughly 10^{27} atoms), then for low enough temperature (and appropriate chemical potential) the probability density would have value almost zero for all configurations except the desired ones – the ones which are the expected crystal (up to a rigid motion). In other words, statistical mechanics doesn't predict precise configurations but an ensemble.

This is a good point to note the discovery, made in 1984 by Dany Schectman *et al*, of materials called "quasicrystals" which are solid but not crystalline. Physicists had gotten used to thinking that solids *had* to be crystalline, though they were aware that this state of affairs did not seem to follow from any known physical law [StO, Ra1]. Quasicrystals were discovered by means of their diffraction patterns (made using wave properties of electrons instead of X-rays, but this is not relevant). The diffraction patterns consisted of lots of "dots" (called Bragg peaks), and some of these patterns were symmetric about a central dot under rotation by $2\pi/10$. Physicists are used to finding such patterns symmetric about a central dot under rotation by $2\pi/6$, or $2\pi/4$, but never by $2\pi/10$. And they were surprised because they knew that no crystal could produce such a pattern. That is, no pattern of atoms which is periodic as in Fig. 2 could produce a diffraction pattern with such a symmetry [HiC; p. 84]. This is a fundamental fact for us, with two aspects.

First, this means the atomic configurations of quasicrystals cannot be periodic as in Fig. 2. And second, there is this connection between the order properties of a pattern (epitomized by the periodicity of periodic patterns), and the symmetry of something associated with the pattern – in this chapter called a diffraction pattern, but given another name in Chapter 3. We will devote the next two chapters to exploring these notions.

In closing this chapter we emphasize an important connection with the previous one. Although everyone's intuitive idea of a crystal is a more or less periodic arrangement of atoms, we just saw above that in physics such a structure is modeled using a probability distribution on a *family* of atomic configurations. At very low temperature that probability distribution is roughly concentrated on the appropriate periodic configuration (and those obtained by Euclidean motions), which shows how the model relates to the intuitive concept. In the chapter on ergodic theory we saw something similar, with translations and invariant integrals living on subshifts. It is no accident that ergodic theory also analyzes a structure using a probability distribution on a family of structures related to the one of interest – after all, historically ergodic theory grew out of the study of matter!

Chapter 3

Order

§1. Spectrum and order

We will be introducing some powerful tools to analyze the order properties of structures like tilings, in the process linking up with our discussion of diffraction in the last chapter. This will involve a roundabout path; we start by embedding the tiling of interest in a family of tilings. Although it may seem odd to introduce many new tilings when we just want information about one specific tiling, this is necessary if we are to use Theorem 1.2 – which gives information about the frequencies of parts of a tiling in terms of an *integral over a space of tilings*.

Assume we want to determine the order properties of a tiling which is an element in some finite type or substitution subshift $X \subset \mathcal{A}^{\mathbb{Z}^d}$. For instance, consider the Morse tiling of Fig. 12, as an element of the substitution subshift $X = X_F$ discussed in Chapter 1, §3. As for an order property, we might want to know: given that there is a tile A at some site s in the Morse tiling, how likely is it there is a tile B at the site s' which is distance 2^{10} to the right of s? Is it more or less likely than if there were a tile B at s?

It will be useful to think of X as a set on which the additive group $\mathbb{Z}^d$ is acting as translations: $(T^t x)_j \equiv x_{j-t}$ for $j \in \mathbb{Z}^d$ and $x \in X$. As in §1.2, we can lift this action to functions on X, for instance the complex continuous functions $C(X)$, by: $T^t f(x) = f(T^{-t} x)$ for $f \in C(X)$. Assuming we have an invariant integral $\mathbb{I}$ on X, invariant with respect to these translations, we now extend the group action to a space a bit larger than $C(X)$.

First we note that $\mathbb{I}$ almost determines an inner product on $C(X)$ by $\langle f, g \rangle \equiv \mathbb{I}(\bar{f} g)$, where $\bar{f}$ is the complex conjugate of f; we say "almost" because $\langle \cdot, \cdot \rangle$ may be missing the property $\langle f, f \rangle = 0 \Rightarrow f = 0$. We now go through a process familiar from group theory, the division of a group by a normal subgroup. Thinking of the complex vector space $C(X)$ as a group under addition, the linear subspace $N \equiv \{ f \in C(X) : \mathbb{I}(|f|^2) = 0 \}$ is a normal subgroup. N is a linear subspace of $C(X)$, since if $f, g \in N$, then $\mathbb{I}(|\alpha f + \beta g|)^2 = \mathbb{I}(|\alpha f|^2) + \mathbb{I}(|\beta g|^2) + \mathbb{I}(\bar{\alpha} \beta \bar{f} g) + \mathbb{I}(\bar{\beta} \alpha \bar{g} f)$, and since $|\mathbb{I}(\bar{g} f)|^2 \leq \mathbb{I}(|g|^2) \mathbb{I}(|f|^2)$ from the Cauchy-Schwarz inequality (Appendix III), we see that $g f \in N$. Since N is a subspace we can divide $C(X)$ by it and get the linear (quotient) space $C(X)/N$ of equivalence classes – two functions are equivalent if their difference is in N – with now a (true) inner product inherited from the almost inner product. (A simple but artificial analogue of this construction is $\mathbb{C}^3$ with the inner product $\langle (a, b, c), (d, e, f) \rangle \equiv \bar{a} d + \bar{b} e$, for which $N = \{ (0, 0, c) : c \in \mathbb{C} \}$. It is a useful exercise to think through the above construction with this example in mind.)

What we have done so far did not use the fact that $\mathbb{I}$ was invariant under translations. We use this now to show that translations are well defined on $C(X)/N$, in other words, that if $f \in N$ then $T^t f \in N$. In fact this is now fairly immediate: $\mathbb{I}(|T^t f|^2) = \mathbb{I}(T^t[|f|^2]) = \mathbb{I}(|f|^2)$. So we can define or lift our representation T of $\mathbb{Z}^d$ from $C(X)$ to $C(X)/N$ by defining the translate of an equivalence class to be the equivalence class of the translate of any of its representatives: $T^t \{ f \} \equiv \{ T^t f \}$. Furthermore, the resulting operators are then "unitary" on the inner product space $C(X)/N$ in the sense that they are linear (easy) and preserve inner products:

$$\langle T^t f, T^t g \rangle = \mathbb{I}(\overline{T^t f} T^t g) = \mathbb{I}(T^t[\bar{f}g]) = \mathbb{I}(\bar{f}g) = \langle f, g \rangle \qquad (3.1)$$

for all f, g. As a last step we complete this inner product space $C(X)/N$ with respect to its norm $\|f\| \equiv \sqrt{\langle f, f \rangle}$, obtaining the Hilbert space we call $\mathcal{H}_{\mathbb{I}}$ (see Appendix III). The operators T^t naturally extend to the completion since they are norm preserving.

It is not hard to go through the above constructions and check that the operators T^t constitute a group representation for $\mathbb{Z}^d$ on $\mathcal{H}_{\mathbb{I}}$, that is: $T^t T^s = T^{t+s}$. (It is even a "continuous" representation in the sense that $\langle f, T^t g \rangle$ is continuous in t for all fixed f, g, since t is a discrete variable; this will be less trivial when we consider translations $t \in \mathbb{R}^d$ in Chapter 4.)

Our next step is to use a convenient Fourier-type decomposition for quantities of the form $\langle f, T^t g \rangle$. This is a fundamental result primarily due to Marshall Stone (see Appendices II, III for notation and concepts).

Theorem 3.1 (Stone's Theorem). *Let $\{T^g\}$ be a continuous unitary representation of the group $G = \mathbb{R}^d$ or $\mathbb{Z}^d$ on a Hilbert space. For each Borel subset B of the space $\hat{G}$ of characters of G there is a (spectral) projection $E(B)$ such that:*

- $E(\emptyset) = 0$;
- $E(\hat{G}) = \mathbb{I}$ *(the identity operator)*;
- $E(B_1 \cap B_2) = E(B_1)E(B_2)$ *(and therefore $E(B_1)E(B_2) = E(B_2)E(B_1)$)*;
- *if $\{B_j\}$ are disjoint, $E(\bigcup_j B_j) = \sum_j E(B_j)$;*
- $E(B) = \inf E(O)$, *the infimum being over all open O containing B.*

With this spectral decomposition we can approximate the operators T^g in the manner: Given $\epsilon > 0$ and $x \in \hat{G}$ (the characters of G), there is *a finite collection of disjoint Borel subsets B_j of $\hat{G}$, containing points g_j, such that $\|T^g - \sum_j g_j E(B_j)\| < \epsilon$.*

(The special case where $G = \mathbb{Z}$ is due to Marshall Stone; various generalizations (beyond $\mathbb{R}^d$ and $\mathbb{Z}^d$) are due to Mark A. Naimark, Warren Ambrose and Roger Godement [RiN; p. 392, Nai; p. 419].) We abbreviate the theorem by writing

$$T^g = \int_{\hat{G}} \hat{g}(g) \, dE(\hat{g}). \tag{3.2}$$

In our case $G = \mathbb{Z}^2$, so the character group $\hat{G}$ is the torus $S^1 \times S^1$, where S^d denotes the unit sphere in d dimensions. The integral representation can then be written:

$$T^{(t_1,t_2)} = \int_0^1 \int_0^1 e^{2\pi i (\theta_1,\theta_2)\cdot(t_1,t_2)} \, dE((\theta_1,\theta_2)). \tag{3.3}$$

The reason we call this a "spectral representation" can be seen by taking a vector f lying in the range of some projection $E(A)$ where A is very small, containing some point $\theta = (\theta_1, \theta_2)$. Then f must be an approximate eigenvector for all the operators $T^{(t_1,t_2)}$, since $T^{(t_1,t_2)} f \approx e^{2\pi i (\theta_1,\theta_2)\cdot(t_1,t_2)} f$, that is, $\|T^{(t_1,t_2)} f - e^{2\pi i (\theta_1,\theta_2)\cdot(t_1,t_2)} f\| \approx 0$. (There may not be any true eigenvector corresponding to (θ_1, θ_2): the approximate eigenvectors corresponding to smaller and smaller sets A may not have a limit in the Hilbert space.)

The red-black checkerboard tilings provide a very simple example. There are only two tilings, depending, say, on the color at the origin in $\mathbb{Z}^2$. So the subshift X consists of two points, x_1, x_2, the invariant integral $\mathbb{I}$ is $\mathbb{I}(f) = [f(x_1) + f(x_2)]/2$, and the Hilbert space $\mathcal{H}_{\mathbb{I}}$ is just the 2-dimensional space $\mathbb{C}^2$ which we think of as column vectors. Since $T^{(t_1,t_2)} x_j = x_j$ if and only if $t_1 + t_2$ is even, we see that the spectral resolution (3.3) reduces to the sum

$$T^{(t_1,t_2)} = e^{2\pi i(0,0)\cdot(t_1,t_2)} E(\{(0,0)\}) + e^{2\pi i(1/2,1/2)\cdot(t_1,t_2)} E(\{(1/2,1/2)\})$$
$$(3.4)$$

with projections concentrated at the two points $(0,0)$ and $(1/2,1/2)$, namely:

$$E(\{(0,0)\}) = \begin{pmatrix} 1/2 & 1/2 \\ 1/2 & 1/2 \end{pmatrix};$$
$$E(\{(1/2,1/2)\}) = \begin{pmatrix} 1/2 & -1/2 \\ -1/2 & 1/2 \end{pmatrix}.$$
$$(3.5)$$

Comparison with this example shows that for any periodic system, such as that associated with Fig. 4, the spectral representation is a finite sum as in (3.4); this must hold because the operators are living on a finite dimensional space. But it is much less easy to determine what can happen for more interesting tilings. One reason this is worth pursuing can be seen from the following two theorems.

Theorem 3.2 [Wal; p. 48]. *The square-like tilings of a subshift* X *exhibit intermediate disorder in the sense of Chapter 1:*

$$\lim_{N\to\infty} (1/N) \sum_{B_N} |\nu(p \cap T^t q) - \nu(p)\nu(q)| = 0 \qquad (3.6)$$

if and only if $E(A)f = 0$ *for any singleton* $A = \{\gamma\}$ *and vector* f, *other than* $A = \{0\}$ *and* f *a constant function, in which case* $E(A)f = f$.

Theorem 3.3 [Wal; p. 66]. *The square-like tilings of a subshift* X *exhibit disorder in the sense of Chapter 1:*

$$\lim_{|t|\to\infty} |\nu(p \cap T^t q) - \nu(p)\nu(q)| = 0 \qquad (3.7)$$

if $dE(\gamma)$ has a "density" in the following sense: for any vector f such that $\langle f, 1\rangle = 0$, $d\langle f, E(\gamma)f\rangle = \rho_f(\gamma)d\gamma$ for some non-negative function ρ_f such that $\int_{\hat{G}} \rho_f(\gamma)\, d\gamma < \infty$.

Both of these results show that when $dE(\gamma)$ has some form of smoothness in γ, the tilings have some form of disorder, and roughly speaking, the more smoothness the more disorder. (We had these results in mind in §1.1.)

To summarize, in this chapter we extended the discussion on order properties of tilings from Chapter 1, linking the notion of order to that of a spectrum of the tiling. This is connected in two ways with the physics in Chapter 2. The idea of analyzing a tiling by embedding it in a family of tilings, with a probability distribution on the family, turns out to be natural from the point of view of the statistical mechanical analysis of structures such as crystals. And second, the spectral analysis of tilings will be shown in the next chapter to be related, by this physics connection, to the X-ray analysis of structures such as crystals. We will now apply what we have learned about order properties to symmetry properties.

Chapter 4

Symmetry

§1. Substitution and rotational symmetry

This book was motivated by the rich mathematics associated with the kite & dart tilings (Fig. 1), in particular the attempt to understand the types of global structures obtainable by local rules of production. By "types" of structures we refer specifically to their order and symmetry properties. So far we have spent most of our time analyzing square-like tilings. This was not an accident – it was on purpose! The reason is that the mathematics of square-like tilings is simpler than that needed for more general tilings such as the kites and darts, and some ideas, notably that of order properties, can be understood more easily in that simpler setting. (There is a close analog in probability theory where the basic ideas, notably independence, are best introduced in the context of discrete random variables, thus avoiding the irrelevant complications of integration.)

In the last chapter we investigated in some depth the order properties obtainable by local rules, and now that we want to move on to the study of symmetry properties, in particular rotational symmetries, we are forced to deal with more general tilings such as the kites and darts, and the pinwheel (Fig. 13). The first thing we must do is upgrade the formalism of Chapter 1 – for instance, the notions of substitution tilings and finite type tilings.

A system of square-like tilings was considered as a subset of a "full shift", the set $\mathcal{A}^{\mathbb{Z}^d}$ carrying a metric and the natural (translation) action of $\mathbb{Z}^d$. Now suppose we have some finite alphabet $\mathcal{A}$ of d-dimensional polyhedra (at fixed positions in $\mathbb{R}^d$), and recall definition 1.9 of $X_{\mathcal{A}}$ as the set of *all* tilings of $\mathbb{R}^d$ by congruent copies of letters of $\mathcal{A}$. We will have to modify the metric we used for the product space $\mathcal{A}^{\mathbb{Z}^d}$ for spaces of tilings such as $X_{\mathcal{A}}$, which are no longer product spaces. The reason is that in the original metric two tilings were close if and only if they coincided perfectly in some large ball around the origin. We can't require that now; if we want to have translations behave continuously it must be the case that two tilings are close if one is a small translate of the other – even though they may not coincide anywhere. And similar considerations apply to rotations.

So we must make some adjustments, though we will try to keep as much of the properties of the previous metric as possible. Heuristically, what we want is that two tilings be close if in a large ball centered at the origin the two have the same polyhedra, in almost the same positions.

What we will use is the following "tiling metric" between tilings x, $x' \in X_{\mathcal{A}}$:

$$m_t(x, \ x') \equiv \sup_{n \geq 1} \frac{1}{n} d_H[B_n(\partial x), B_n(\partial x')], \qquad (4.1)$$

where $d_H[A, B]$ is the Hausdorff metric between two compact subsets A, B of $\mathbb{R}^d$, defined by

$$d_H[A, B] = \max\{\tilde{d}(A, B), \tilde{d}(B, A)\}, \qquad (4.2)$$

where

$$\tilde{d}(A, B) = \sup_{a \in A} \inf_{b \in B} |a - b|, \qquad (4.3)$$

and $B_N(\partial x)$ denotes the union of those portions, of the boundaries of tiles in the tiling x, which are contained in the closed ball $\bar{B}_N \subset \mathbb{R}^d$ of volume N centered at the origin. (Compare this with the metric on subshifts introduced in §2 of Chapter 1.) It is not hard to show that $X_{\mathcal{A}}$ is compact in this metric [RaW]. We think of $X_{\mathcal{A}}$ as the analog of a finite type subshift, and will sometimes be considering subsets $X \subset X_{\mathcal{A}}$ produced by a substitution rule, inheriting the metric.

As our first example we take the kite & dart tilings made from the quadrilaterals of Fig. 16. Consider the space $X_{\mathcal{A}_{16}}$ of all tilings made from these letters. Since the two letters fit together to form a rhombus (see Fig. 16 and imagine the dart shifted to sit atop the kite), translates of which can then tile the plane, $X_{\mathcal{A}_{16}}$ contains periodic tilings besides the ones in which we are really interested. Imitating Chapter 3, we define the (substitution) kite & dart tilings $X_{k\&d}$ as the subset of $X_{\mathcal{A}_{16}}$ consisting of those tilings x in which every word is a subword of one of those special ones made by iterating the substitution function of Fig. 17. As a subset of $X_{\mathcal{A}_{16}}$, $X_{k\&d}$ inherits a metric and (since a Euclidean image of a tiling in $X_{k\&d}$ is again in $X_{k\&d}$) a continuous representation of the Euclidean group $\mathcal{E}^2$ of the plane.

These tilings have been inspirational; in particular they inspired the mathematics discussed in this book! (For an introduction to the impact the tilings have had on a variety of disciplines, we recommend [Sen].) One of their pretty features is the pentagonal regions that one sees throughout the tilings. Not only are there the five-sided stars made of five kites, and the decagons made of five darts (see Fig. 1), but because of the substitution there must be regions of arbitrarily large size with such local 5-fold symmetry. Although there is much discussion in the literature of this "approximate 5-fold symmetry", as noted in the Introduction we will be examining a certain type of *10-fold* symmetry of these tilings, which will be seen to originate in the fact that in the substitution function of Fig. 17 there are (small) kites and darts rotated with respect to one another by $2\pi/10$.

Before we explore properties of these tilings we consider the finite
type version made from the two letters of Fig. 18. That is, if $\mathcal{A}_{18}$ is the
alphabet of these two letters, we consider the set $X_{\mathcal{A}_{18}}$ of *all* tilings
with these letters. As noted in §4 of Chapter 1, there is a natural one-
to-one correspondence between $X_{\mathcal{A}_{18}}$ and $X_{k\&d}$ obtained by ignoring
the small bumps and dents in the letters of $\mathcal{A}_{18}$. In fact this map and
its inverse are easily seen to be continuous, and furthermore they each
carry the action of $\mathcal{E}^2$ on one space to its action on the other space,
so for all practical purposes $X_{\mathcal{A}_{18}}$ and $X_{k\&d}$ are interchangeable.

At this point it is convenient to discuss some material which shows
why substitution is so useful. In particular it supplies the promised
proof of Lemma 1.9.

Lemma 4.1. *Let X_F be the substitution tiling system associated with
the substitution function $F = E_\gamma \circ \tilde{F}$. If each tiling $x \in X_F$ is uniquely
decomposable into a tiling x_F of letters of level n for all n (note that
x_F is not in X_F, but only because its tiles are not the right size), then
the map $T^F : x \in X_F \to E_{1/\gamma}(x_F) \in X_F$ is a representation of the
similarity $E_{1/\gamma} : p \in \mathbb{R}^d \to p/\gamma \in \mathbb{R}^d$; that is, T^F has the correct
group relations with the natural representation of the Euclidean group
on X_F. Furthermore, for any $x \in X_F$, $T^t x = x$ implies $t = 0$.*

Before we get to the proof we make a few explanatory comments.
Whenever you have a substitution tiling you can, by Lemma 1.4,
think of it as a tiling by larger size letters, letters of level k for any
k; see Fig. 14. Lemma 4.1 deals with the slightly special case where
this decomposition into higher-level letters is unique. (Recall the
discussion after Lemma 1.4 of the periodic checkerboard tiling.) And
the conclusion is two-fold. First, the uniqueness provides a natural
map from tilings (by level 0 tiles) to tilings (by level 0 tiles) – we
take the map from tilings by level 0 tiles to tilings by level 1 tiles
and shrink the level 1 tiles about the origin so they become level 0
tiles. This (composite) map is then claimed by this lemma to be
a representation of a similarity of the plane, in that this map has

the same group relations with the Euclidean motions of the tilings in X_F as does the corresponding similarity of Euclidean space with the Euclidean motions of points in Euclidean space $\mathbb{R}^d$. And second, it is claimed that this group property implies that no substitution tiling has any translational symmetry, as in Lemma 1.9.

Proof of Lemma 4.1. The first part of the lemma follows easily from the form of T^F, so we will only discuss the last claim. The proof will be by contradiction. So assume for some tiling x_0 that $T^{t_0}x_0 = x_0$. Since T^F represents the similarity $E_{1/\gamma}$, for any translation t

$$(T^F)T^t(T^F)^{-1} = T^{t/\gamma}. \tag{4.4}$$

Then for all n we have $(T^F)^n T^t = T^{t/\gamma^n}(T^F)^n$. Therefore $(T^F)^n x_0 = T^{t_0/\gamma^n}(T^F)^n x_0$. In other words the tiling $(T^F)^n x_0 \in X_F$ is invariant under the translation t_0/γ^n. Since for large n this is smaller than any tile in the tiling, we must have $t_0 = 0$. $\qquad\square$

This lemma shows two things: the abstract fact that substitution tiling systems carry interesting representations of similarities, and the consequence that such tilings are automatically nonperiodic.

We now get back to our analysis of the rotational symmetries of the kite & dart tilings. To understand them we begin by observing that $X_{k\&d}$ is unwieldy – it decomposes into more readily understood subsets. In fact it decomposes into a family $\{X_{k\&d}^\alpha : \alpha \in [0, 2\pi/10)\}$ of closed subsets related to one another in a simple way: the tilings in any one of these subsets, say $X_{k\&d}^{\alpha_0}$, are merely the rotations (about any point) of those of any other one $X_{k\&d}^\alpha$, in fact with rotation by $\alpha - \alpha_0$. So for kites and darts it suffices to understand any one of these subsets $X_{k\&d}^\alpha$. Now such a set can be investigated just as we did the substitution subshifts in §3 of Chapter 1. In particular it follows from the same simple conditions (Lemma 1.5) that there is one and only one invariant integral $\mathbb{I}_\alpha$ on $X_{k\&d}^\alpha$ which is invariant under translations. The new idea is: what happens to $X_{k\&d}^\alpha$ when we rotate about some fixed point by $2\pi/10$? (The answer is that $X_{k\&d}^\alpha$ gets mapped onto itself by such a rotation. Think about this!) So consider the integral $T^{R(2\pi/10)}\mathbb{I}_\alpha$ obtained from $\mathbb{I}_\alpha$ by such a rotation: i.e., $T^{R(2\pi/10)}\mathbb{I}_\alpha(f) \equiv \mathbb{I}_\alpha(T^{R(-2\pi/10)}f)$, with $R(\theta)$ representing rotation by θ about the origin. It is easy to check that $T^{R(2\pi/10)}\mathbb{I}_\alpha$ is also a translation invariant integral on $X^\alpha_{k\&d}$. And since we know that $\mathbb{I}_\alpha$ is the *only* such beast, we must have $T^{R(2\pi/10)}\mathbb{I}_\alpha = \mathbb{I}_\alpha$. In other words, the integral $\mathbb{I}_\alpha$ is not only translation invariant but also invariant under rotation (about any point) by $2\pi/10$. (It is useful to realize how simply this followed from the uniqueness of $\mathbb{I}_\alpha$ among translation invariant integrals.)

We now want to re-interpret this symmetry of the integral $\mathbb{I}_\alpha$ as a symmetry of individual tilings. We will need to use the following analog of Theorem 1.2. (It might be useful to review §2 of Chapter 1, in particular the connection between invariant integrals and the frequencies in tilings.)

Theorem 4.2. *Suppose there is a continuous representation T of the group $\mathbb{R}^d$ on the compact metric space X, and $\mathbb{I}$ is an invariant integral on X. Then the following three conditions are equivalent:*

i) $\mathbb{I}$ is the only invariant integral on X;

ii) for every continuous function f on X and $x \in X$,

$$|\frac{1}{N}\int_{B_N}[T^t f(x) - \mathbb{I}(f)]\,dt| \underset{N\to\infty}{\longrightarrow} 0; \qquad (4.5)$$

iii) for every continuous function f on X,

$$\sup_{x\in X}|\frac{1}{N}\int_{B_N}[T^t f(x) - \mathbb{I}(f)]\,dt| \underset{N\to\infty}{\longrightarrow} 0. \qquad (4.6)$$

Corollary 4.3 [Pet; p. 271]. $\mathbb{I}$ *is the only invariant integral on X if and only if for any open subset S of X for which the indicator functions χ_S and $\chi_{\bar{S}}$ of S and its closure $\bar{S}$ satisfy $\mathbb{I}(\chi_S) = \mathbb{I}(\chi_{\bar{S}})$ we* *have:*

$$|\frac{1}{N} \int_{B_N} [T^t \chi_S(x) - \mathbb{I}(\chi_S)] \, dt| \xrightarrow[N \to \infty]{} 0 \qquad (4.7)$$

for every $x \in X$.

We apply these ideas to a space X of tilings by choosing an $S \subset X$ of the following sort. Fix some finite word $\mathcal{P}$ in a tiling of X, and metric m_E on the Euclidean group $\mathcal{E}^d$. Then pick some $\epsilon > 0$ and define $S(\mathcal{P})$ as the set of all tilings x such that $T^g x$ has the word $\mathcal{P}$ in the same place for *some Euclidean motion g* which is ϵ-close to the identity e of $\mathcal{E}^d$, that is, $m_E(g, e) < \epsilon$. It follows from the continuity of the representation of $\mathcal{E}^d$ that $S(\mathcal{P})$ is an open subset of X, and one can in fact check that $S = S(\mathcal{P})$ satisfies the conditions of the above corollary.

Note that for $X = X^\alpha_{k \& d}$ the only Euclidean motions near the identity which map X into itself are pure translations. So we can now reproduce the argument of Chapter 1 to interpret the value of $\mathbb{I}(\chi_{S(\mathcal{P})})$ as the *frequency* with which the pattern $\mathcal{P}$ appears (in its original orientation) in *any* tiling in X. As in the simpler case of subshifts, knowing $\mathbb{I}(\chi_{S(\mathcal{P})})$ for all $S(\mathcal{P})$ allows one to reproduce $\mathbb{I}$, in the sense that the values $\mathbb{I}(f)$ for general f are determined by the values for the special f of the form $\chi_{S(\mathcal{P})}$. Putting this together with our new result that $\mathbb{I}$ is invariant under rotation by $2\pi/10$, we reinterpret this as saying that for every tiling x the frequency of any finite word $\mathcal{P}$ is the same as it is in the rotated tiling $T^{R(2\pi/10)} x$.

As noted in the Introduction, this is a fundamentally new idea: that a pattern could have a Euclidean symmetry, namely a rotational symmetry, in a sense slightly weaker than the usual one whereby the whole pattern itself is unchanged under the motion; in this weaker form only the frequencies of parts of the pattern are unchanged under the motion, not the global pattern itself. We will see the value of this new form of symmetry throughout the remainder of the book.

In our next example we will see this statistical form of symmetry displayed in a way manifestly impossible for the traditional form – rotational symmetry of a polygonal tiling by *all* angles! (In fact it was this example ([Ra4]) which clarified the important distinction between the approximate 5-fold rotational symmetry and statistical 10-fold symmetry of the kite & dart tilings). Consider again the pinwheel tiling of Fig. 13. The substitution version of the pinwheel is made from a $1, 2, \sqrt{5}$ right triangle by the substitution function of Fig. 30, and is used in the usual way to produce a space X_{pin} of pinwheel tilings. (There is also a finite type version of the pinwheel [Ra2].) We want to focus first on the rotational symmetries of these tilings, and this requires overcoming a technical problem.

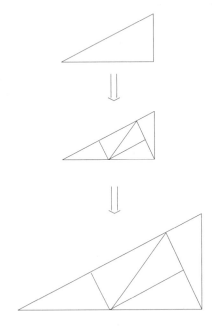

Figure 30. The pinwheel substitution

For the kite & dart tilings $X_{k\&d}$ we used the simplification that $X_{k\&d}$ decomposes into the subsets $X_{k\&d}^\alpha$. Technically this was advantageous since the sets $S(\mathcal{P})$ could be thought of as tilings containing small *translations* of the word $\mathcal{P}$, not small rotations. For the pinwheel we have to work in the more general setting. The analysis of the pinwheel tiling has some similarity to the proof of Lemma 1.5, which dealt with square-like tilings. It is useful to repeat the analysis in this setting to see how rotations enter the mathematics, a major concern in this book.

Theorem 4.4. *On the space X_{pin} of pinwheel tilings there is one and only one invariant integral.*

Proof. (It may be useful to review Appendices II and III.) From the above, this follows if and only if each finite word $\mathcal{P}$ appears in each tiling x with the same frequency *in each fixed range of orientations*. For instance, picking some finite clump of tiles in the pinwheel tiling of Fig. 13, we want to prove that the frequency with which that clump appears in an orientation rotated from the defining orientation by an angle in the interval $(0, \pi/20)$ is the *same* as the frequency with which it appears rotated by an angle in the interval $(\pi/15, 7\pi/60)$. (Unlike the situation for the kites and darts, here we cannot usefully attribute a frequency to a word at fixed orientation – we must attach it to a range of orientations. We will refer to a "frequency density" of words, to emphasize the dependence of the frequency on orientation.) We will show this in two parts: we will first prove that, neglecting their orientations, each word appears with the same frequency in each tiling – this is very similar to what we did in Chapter 1 – and then we will show the independence of orientation.

We will take for our alphabet $\mathcal{A}$ a 1, 2, $\sqrt{5}$ right triangle and its reflection, so the only Euclidean motions we need are translation and rotation. In Chapter 1 we used a matrix method to keep track of how many times a given letter appears in a letter of level p, that is, after the substitution is iterated p times. We now need to keep track of the absolute orientations of a letter, not just how many times it appears. To do this we will incorporate in our machinery the irreducible unitary representations of the rotation group $SO(d)$. (Even though we are ostensibly discussing the pinwheel tilings, we will sometimes use general notation to emphasize how the method generalizes.) In the case of the pinwheel $d = 2$, $SO(d)$ is just the Abelian group $[0, 2\pi)$ under addition modulo 2π, and its irreducible representations are the characters $f_m : \alpha \in [0, 2\pi) \rightarrow e^{im\alpha} \in \mathbb{C}$, $m \in \mathbb{Z}$ (Appendix II). To include these representations we modify the previous matrices. Instead of one matrix we now have a *family* of $K \times K$ matrices, where K is the size of the alphabet ($K = 2$ for the pinwheel), labeled by the irreducible representations of $SO(d)$ (labeled by $m \in \mathbb{Z}$ for the pinwheel); the j, k entry $M(m)_{j,k}$ is:

$$M(m)_{jk} = \sum_{\ell} f_m[A_{jk}(\ell)], \qquad (4.8)$$

where $A_{jk}(\ell) \in SO(d)$ is the relative orientation (with respect to the standard letter) of the ℓth copy of the type j letter in the substitution expansion of the type k letter. For the pinwheel, using $s \equiv \arctan(1/2)$, $M(m)$ is:

$$\begin{pmatrix} e^{-ims} + e^{-im(s+\pi)} & 2e^{-im(s+\pi)} + e^{-im(s+3\pi/2)} \\ 2e^{im(s+\pi)} + e^{im(s+3\pi/2)} & e^{ims} + e^{im(s+\pi)} \end{pmatrix}. \qquad (4.9)$$

(Check this against Fig. 30.)

In light of the above corollary we want to prove that every word has the same frequency density in all tilings and at all orientations. It is convenient to prove first, using arbitrary orientations, that the frequencies of words are independent of the tiling. As for subshifts, one can show that it suffices to consider a word consisting of a single letter, and regions which are letters of increasing level: $F_p^r(a)$ for arbitrary a and r. If a is type k, the number of type j letters in

$F_p^r(a)$ is $[M(0)^r]_{jk}$ where

$$M(0) = \begin{pmatrix} 2 & 3 \\ 3 & 2 \end{pmatrix}. \tag{4.10}$$

Just as in §1.3 we can apply the Perron-Frobenius theorem to $M(0)$ to show that the words appear with the same frequency in every tiling, neglecting orientation. The new problem is to refine the argument to handle the case where we only consider a word in some fixed range of orientations; we want to show that, given two equal length open intervals of orientations (say $(0, \pi/20)$ and $(\pi/15, 7\pi/60)$), the frequency of any word with orientation in one interval is the same as its frequency with orientation in the other interval, for any pinwheel tiling. We compute frequencies by counting in larger and larger regions of a tiling and dividing by the area. And as in the above argument using the Perron-Frobenius theorem, we will restrict attention to regions which are occupied by a letter of some (increasingly large) level. So in fact given an initial letter, we have a sequence of finite sequences – the list of orientations of some letter type in ever higher-level letters. What we need to prove therefore is what is usually called the "uniform distribution" of the orientations – that as you add more orientations to your list, the frequencies ν_B and $\nu_{B'}$, of hits in any two open intervals B and B' of equal size, have the same limit.

There is a well known "Weyl criterion" for proving uniform distribution:

Theorem 4.5 (Weyl) [KuN; p. 226]. $\{x_j\} \in SO(d)$ *is uniformly distributed if and only if for every continuous nontrivial irreducible representation f of $SO(d)$,*

$$\lim_{N \to \infty} (1/N) \sum_{j=1}^{N} f(x_j) = 0. \tag{4.11}$$

This criterion is best known in the special case $d = 2$ (which we need for the pinwheel), where it says that a sequence $\{x_j\}$ on the unit circle is uniformly distributed if and only if for $m \neq 0$:

$$\frac{\sum_{k=1}^{K} e^{imx_k}}{K} \xrightarrow[K\to\infty]{} 0. \tag{4.12}$$

But in a sense the more general result is more revealing as it says that to understand whether or not a sequence on $SO(d)$ is uniformly distributed one can linearize the problem – analyze it in certain representations of the group $SO(d)$, effectively reducing the question to the behavior of certain matrices instead of points in the complicated space $SO(d)$.

Getting back to our proof, and noting how the orientations appear in $F_p^k(a)$, in order to prove that the orientations are uniformly distributed we need to show

$$\frac{[M(m)^p]_{jk}}{[M(0)^p]_{jk}} \xrightarrow[p\to\infty]{} 0, \quad \text{for all } m \neq 0. \tag{4.13}$$

To complete the proof of Theorem 4.4 we first note that for all j, k and m, $|M(m)_{jk}| \leq |M(0)_{jk}|$. Assume that for some j, k, and some p, $|M(m)_{jk}^p| < |M(0)_{jk}^p|$ and that $M(0)$ is primitive (i.e., all coordinates are nonnegative and some power has all coordinates strictly positive), both of which are true for the pinwheel. We will show by contradiction that the spectral radius $\gamma(m)$ of $M(m)$ is less than that of $M(0)$, λ, for all m. Then $\gamma(m)^p$ is the spectral radius of $M(m)^p$ [BaN; p. 322], with corresponding eigenvector $\psi(m)$. Using the subscript $+$ to replace the components of column vectors and matrices with their absolute values, we have

$$|\gamma(m)^p|\psi_+(m) \leq [M(m)^p]_+\psi_+(m) \leq [M(0)^p]\psi_+(m) \leq \lambda^p\psi_+(m). \tag{4.14}$$

Therefore $|\gamma(m)| \leq \lambda$. And if $|\gamma(m)| = \lambda$, then $\psi_+(m)$ must be the Perron-Frobenius eigenvector of $M(0)^p$, and $M(0)^p\psi_+(m) \ =$

$[M(m)^p]_+ \psi_+(m)$, so $[M(m)^p]_+ = M(0)^p$, a contradiction. This shows $\gamma(m) < \lambda$, and this proves our claim about the spectral radii of $M(m)$ and $M(0)$. From this, we note that $|M(m)_{jk}^r| \leq \|M(m)^r\|$ for all j, k, and $\|M(m)^r\|^{1/r} \longrightarrow \gamma(m)$. So

$$\frac{|[M(m)^r]_{jk}|}{[M(0)^r]_{jk}} \leq \frac{\|M(m)^r\|}{[M(0)^r]_{jk}} \equiv a_r. \tag{4.15}$$

We know $a_r^{1/r} \longrightarrow \gamma(m)/\lambda < 1$, so $a_r \longrightarrow 0$, which completes our proof that there is only one invariant integral on the pinwheels. $\square$

In the next section we will make a connection between the statistical rotational symmetry of tilings such as the kites and darts, or pinwheels, and the atomic structure of quasicrystals.

§2. Spectrum and X-rays

Statistical rotational symmetry is concerned with the symmetry of an integral on a space of structures such as tilings. We want to connect that notion with a model of the X-ray diffraction of some configurations of scatterers, as in Chapter 2. We will do this by first transferring some of what we have developed in Chapters 3 and 4 for tilings to configurations of scatterers.

When we analyzed diffraction in Chapter 2 we considered the realistic situation of diffracting off a finite collection of scatterers. In order to get clean results we will need to consider limits of those results as the number of scatterers goes to infinity. Specifically, we will imagine we have an infinite configuration of scatterers, and will compute what the diffraction would be off a sequence of larger and larger subsets – say, the scatterers in the sequence of balls B_N of volume N centered at the origin.

We need some formalism. Fix any two constants $0 < a < b$ and let $\mathcal{X}^d$ be the collection of all infinite sets x of points in $\mathbb{R}^d$ satisfying the following two conditions: no ball of radius a in $\mathbb{R}^d$ contains more than 1 point of x; every ball of radius b in $\mathbb{R}^d$ contains at least 1 point of x. On $\mathcal{X}^d$ we put a "configuration metric" m_c just as we did for tiling systems: the distance between two configurations is given by

$$m_c(x,\ x') = \sup_N \frac{1}{N} d_H[B_N \cap x, B_N \cap x'], \qquad (4.16)$$

where we use the Hausdorff metric d_H as we did in (4.1). As with tiling spaces, we can show that $\mathcal{X}^d$ is compact and that the natural representation of Euclidean motions on $\mathcal{X}^d$ (for which we will use the same notation as for tilings) is continuous. For diffraction purposes let's assume we have some one configuration $\tilde{x} \in \mathcal{X}^d$ of interest and define $\tilde{X} \subset \mathcal{X}^d$ as the closure of the orbit $\{T^t\tilde{x} : t \in \mathbb{R}^d\}$ of $\tilde{x}$ under translations. We will assume that there is only one invariant integral $\mathbb{I}$ on $\tilde{X}$ (see [Ra1] for physical justification of this assumption). We showed in (2.8) that the scattering intensity of a density ρ_V of scatterers in volume V is the Fourier transform of the autocorrelation of the density. Given a configuration $\tilde{x}$ as above, we now construct densities ρ_V for use in our diffraction formulas of Chapter 2.

For $x \in \tilde{X}$ let $\omega(x)$ be any point in x closest to the origin. Fix any $0 < \epsilon < a/2$ and some nonnegative real continuous function f on $\mathbb{R}^d$ such that $f(0) > 0$ and $f(s) = 0$ for $|s| > \epsilon$. (f will later be interpreted as the density of an individual scatterer, and at some point we will assume it is invariant under rotation about the origin.) Define the real function $\tilde{f}$ on $\tilde{X}$ by

$$\tilde{f}(x) = f[-\omega(x)]. \qquad (4.17)$$

(Note that $\tilde{f}$ is well-defined and continuous.) Consider the function $G(x) \equiv [T^u \tilde{f}(x)]\tilde{f}(x)$ on $\tilde{X}$. By Theorem 4.2,

$$\frac{1}{|V|} \int_V [T^{u+u'}\tilde{f}(x)]T^{u'}\tilde{f}(x)\,du' \xrightarrow[|V|\to\infty]{} \mathbb{I}([T^u\tilde{f}]\tilde{f}) \qquad (4.18)$$

for every $x \in \tilde{X}$. Now we generate an electron density in V corresponding to x given by:

$$\rho_V(u) = T^u\tilde{f}(x) = f[-\omega(T^{-u}x)]. \qquad (4.19)$$

The left hand side of (4.18) is just the autocorrelation per unit volume $A_V(u)/|V|$ (see (2.8)), and the right hand side of (4.18), thought of as the inner product $\langle T^u \tilde{f}, \tilde{f} \rangle$, has spectral resolution:

$$\mathbb{I}([T^u \tilde{f}]\tilde{f}) = \int_{\mathbb{R}^3} e^{2\pi i u' \cdot u} \, d\langle \tilde{f}, E(u')\tilde{f} \rangle. \tag{4.20}$$

So from (2.6) – (2.8) (assuming $E_0 = 1$ for simplicity) we see that (in some sense):

$$\frac{I_V(s)\, ds}{|V|} \xrightarrow[|V| \to \infty]{} d\langle \tilde{f}, E(s)\tilde{f} \rangle, \tag{4.21}$$

a formula due to Steven Dworkin [Dwo].

Formula (4.21) gives us a connection between the scattering intensity per unit volume off a configuration $\tilde{x}$ and the spectral projections $\{E(s)\}$ of the translations on $\tilde{X}$ coming from Stone's Theorem 3.1. So the (symmetry) properties of the scattering intensity can be related to the properties of the $\{E(s)\}$, which we examine next.

First let's reconsider the red-black checkerboard tilings discussed in Chapter 3. In that chapter we considered the tilings as a subshift, so the only translations available were $\mathbb{Z}^2$. This time we will allow translations from $\mathbb{R}^2$, as well as rotations. So the space of tilings becomes a torus $\mathcal{T} \equiv \{\mathcal{T}^\alpha : 0 \le \alpha < \pi/2\}$ whose elements are themselves tori $\mathcal{T}^\alpha \equiv \{\theta^\alpha = (\theta_1^\alpha, \theta_2^\alpha) : 0 \le \theta_j^\alpha < 2\}$, one for each orientation α, $0 \le \alpha < \pi/2$. We take as our translation invariant integral $\mathbb{I}$ the usual normalized area on one of these tori, say, $\mathcal{T}'$. As before, translations are represented unitarily on the Hilbert space $\mathcal{H}_\mathbb{I}$ and we use the notation T^t for these translation operators. Note that

$$(T^t\theta')_k = \theta'_k + t_k \,(\mathrm{mod}\ 2). \tag{4.22}$$

If we define, for all $j \in (\mathbb{Z}/2)^2$,

$$\psi_j(\theta') \equiv \exp(2\pi i\, j \cdot \theta'), \tag{4.23}$$

then $\psi_j \in \mathcal{H}_\mathbb{I}$, $\|\psi_j\|^2 = 1$, and $T^t\psi_j(\theta') = \exp[-2\pi i\, j \cdot t]\psi_j(\theta')$. So the ψ_j are normalized eigenvectors of the T^t. In fact they form a (well known) orthonormal basis of $\mathcal{H}_{\mathbb{I}}$. Therefore for any $\psi \in \mathcal{H}$, $\psi = \sum_{j \in (\mathbb{Z}/2)^2} a_j \psi_j$, where $a_j = \langle \psi_j, \psi \rangle$. So

$$
\begin{aligned}
T^t \psi &= \sum_{j \in (\mathbb{Z}/2)^2} a_j \exp(-2\pi i\, j \cdot t)\, \psi_j \\
&= [\sum_{j \in (\mathbb{Z}/2)^2} \exp(-2\pi i\, j \cdot t) P_j] \psi,
\end{aligned}
\tag{4.24}
$$

where P_j is the 1-dimensional projection $P_j \psi = \langle \psi_j, \psi \rangle \psi_j$. So (4.24) is Stone's formula (3.2), with $dE(\gamma)$ concentrated on $(\mathbb{Z}/2)^2$.

We need to understand how the rotational symmetries of this spectral decomposition come about. We noted in 4.22) how the translations operate on each piece $\mathcal{T}^\alpha$ of our space $\mathcal{T}$ of tilings. Rotations are more complicated: rotation (about any fixed axis) by an angle ξ maps $\mathcal{T}^\alpha$ onto $\mathcal{T}^\beta$ where $\beta = \alpha + \xi \pmod{\pi/2}$. In particular, only rotation by a multiple of $\pi/2$ maps a $\mathcal{T}^\alpha$ into itself. Consider such a rotation R. Every tiling in $\mathcal{T}^\alpha$ has points about which the tiling is unchanged by rotation by $\pi/2$ – different points for different tilings. Recall the interpretation of the integral $\mathbb{I}$ as a list of the frequencies with which patterns appear, at given orientation, in the tilings in $\mathcal{T}^\alpha$. Because of this rotational symmetry of the tilings, any pattern would appear in a tiling with the *same* frequency in one orientation as it would in the orientation rotated by $\pi/2$, and this implies that $\mathbb{I}$ is invariant under rotation (about any point) by $\pi/2$. Then the same argument that we used for translations applies to show that such a rotation is unitarily implemented in the Hilbert space $\mathcal{H}_{\mathbb{I}}$, say by the operator R. And finally, we see that for subsets $A \subset \mathcal{T}^\alpha$ the spectral projection $E[R(A)]$ associated to the rotated set $R(A)$ is related to the projection $E(A)$ for the unrotated set by

$$
E[R(A)] = RE(A)R^{-1}.
\tag{4.25}
$$

Now consider some family $X^\alpha_{k\&d}$ of kite & dart tilings. This time, although none of the tilings is invariant under rotation by $2\pi/10$, as we showed earlier in this chapter, it's still true that the translation invariant integral $\mathbb{I}_\alpha$ is invariant under such a rotation. So again we have that the spectral projections satisfy the analog of (4.25), with R now denoting rotation (about any fixed point) by $2\pi/10$.

The difference in the two situations is important, so we will review the arguments. For the checkerboard tilings we used the fact that each tiling had some point about which it was invariant under rotation by $\pi/2$ (the point varying with the tiling). From this it followed that the translation invariant integral was invariant under such a rotation about any point, and this implied that such a rotation was unitarily implemented in the appropriate Hilbert space, and then that the spectral projections for translations satisfied (4.25). For the kite & dart tilings the argument starts differently – now the invariance of the integral under rotation (by $2\pi/10$) is obtained by the substitution rule, not by the invariance of the tilings – but the rest of the argument leading to (4.25) is the same as for checkerboard tilings.

The difference in argument was significant historically [StO, Sen]. Physicists have had a lot of experience analyzing solids, and had gotten used to the situation that solids all seemed to be crystalline (at the microscopic level), that is, had an atomic configuration roughly of the form of Fig. 2. Part of the reasoning was that diffraction off solids always gave results consistent with such an atomic configuration. Then when diffraction patterns with 10-fold rotational symmetries were discovered in 1984, it was quite surprising. After all – they knew from the classification of the crystallographic groups [NiS; p. 149] that no crystal could produce a diffraction pattern with such a symmetry. On the other hand they couldn't understand how *any* configuration could produce a diffraction pattern with such a symmetry. The resolution seems to be that the diffraction symmetry, which is just a spectral symmetry according to the analysis of Dworkin (4.21), is associated with a *statistical rotational symmetry* of the configuration rather than the more common situation in which the configuration itself is rotation invariant; this leads to a rotational symmetry of the diffraction from the invariance under rotation (about the origin) if, as is natural,

the density $\tilde{f}$ in (4.21) is rotation symmetric about the origin – i.e.,
$R\tilde{f} = \tilde{f}$.

§3. 3-dimensional models

For the pinwheel tilings we showed in §4.1 how an irrational rotation
in the substitution (by $2\arctan(1/2)$) leads to the invariance under
all rotations of any translation invariant integral on the system. This
turned out to be quite significant, as it was the mechanism which
led to the discovery ([Ra4])of the new phenomenon of the statistical
rotational symmetry of hierarchical tilings such as the kite & dart,
and pinwheel.

To get a comparable effect in a 3-dimensional tiling we could look
for a substitution which uses irrational rotations about different axes.
Instead we will construct examples which achieve the desired goal
(full rotational symmetry of the translation invariant integrals) by a
different means – the *noncommutativity* of rotations in three dimensions. This will require our first use of nontrivial parts of algebra,
namely some commutative ring theory. The justification for this is
not just to play with another part of mathematics, as enjoyable as
that is, but that this path gives insight into an essentially new feature
of tiling in three dimensions, the noncommutativity of the rotations.
Much more important, it actually leads ([CoR, RS1, RS2]) to new,
significant mathematics, the basic relationships between simple pairs
of rotations in space. This is our best example of how the interdisciplinary exploration of tilings has led to unexpected results of real
interest in seemingly unrelated parts of mathematics.

Our first example of a 3-dimensional tiling, made in collaboration
with John Conway [CoR], is called "quaquaversal". The alphabet
consists of a single triangular prism, with triangular legs of lengths
1 and $\sqrt{3}$, and depth 1. The substitution function $F = E_\gamma \circ \tilde{F}$ first
decomposes the prism into eight small prisms as shown in Fig. 31,

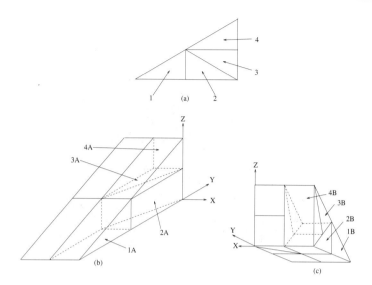

Figure 31. (a) Decomposing a triangular face of the quaquaversal tile; (b,c) two views of the decomposition of the quaquaversal tile

and then expands by a linear stretch by the factor $\gamma = 2$. Part of a quaquaversal tiling is shown in Fig. 32.

In §1 of this chapter we used Weyl's criterion (Theorem 4.5) to show the uniform distribution of orientations in the pinwheel. Generalizing the argument to three dimensional tilings, such as the quaquaversal, is not hard; we just have to allow for the fact that beyond two dimensions the irreducible representations of the rotation group need no longer be one dimensional. (Recall the use of uniform distribution in the proof of Theorem 4.4.)

Theorem 4.6. *The orientations of the tiles in a quaquaversal tiling are uniformly distributed in* $SO(3)$.

Proof. We must show that $(1/N) \sum_{n=1}^{N} f_{ij}(g_n) \to 0$ for every matrix element f_{ij} of every continuous irreducible unitary representation f

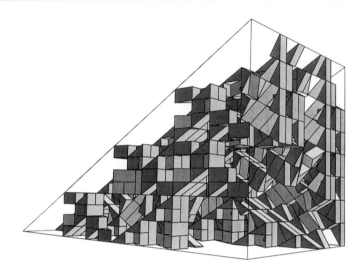

Figure 32. Part of a quaquaversal tiling

of $SO(3)$ other than the trivial one. Using the methods of [Ra3] we can restrict attention to $N = 8^k$, corresponding to k iterates of the substitution. In such a structure the orientations of the tile are the summands in the element $[g'_1 + \cdots + g'_8]^k$ of the group algebra of $SO(3)$ (see Appendix II), where $g'_1, \cdots, g'_8$ are the eight orientations of the small tiles with respect to that of their decomposed parent. (Specifically, with respect to the x, y, z axes of Figs. 31b,c and using the notation e for the identity and R_x^θ for rotation by angle θ about the x axis, the eight orientations of tiles $1A, \cdots, 4A$ and $1B, \cdots, 4B$ are, respectively: $e, R_y^{2\pi/4} R_z^{2\pi/2}, R_y^{6\pi/4}, e$, and $e, R_z^{2\pi/2}, R_x^{2\pi/3} R_y^{2\pi/2}, R_x^{2\pi/3}$.)

Now, using the operator norm $\| \ \|$,

$$|([f(g'_1) + \cdots + f(g'_8)]^k)_{ij}| \leq \|[f(g'_1) + \cdots + f(g'_8)]^k\|. \qquad (4.27)$$

So

$$([f(g_1') + \cdots + f(g_8')]^k)_{ij}/8^k \longrightarrow 0 \qquad (4.28)$$

as $k \to \infty$ if

$$\||[f(g_1') + \cdots + f(g_8')]^k\||/8^k = (\||[f(g_1') + \cdots + f(g_8')]^k\||^{1/k}/8)^k \longrightarrow 0. \qquad (4.29)$$

But $\||[f(g_1') + \cdots + f(g_8')]^k\||^{1/k}$ has as its limit the spectral radius of $f(g_1') + \cdots + f(g_8')$ (see Appendix III), so the limit in (4.28) can only be nonzero if that spectral radius is 8; it certainly cannot be larger than 8 since the norm cannot be larger than 8. We prove it is not 8 by contradiction as follows. Assuming the spectral radius is 8, there is a (unit length) eigenvector ϕ of $f(g_1') + \cdots + f(g_8')$ with eigenvalue of absolute value 8, and since each $f(g_j')\phi$ is of unit length, they must be the same vector for all $j = 1, \cdots, 8$, namely the same multiple of ϕ. But then ϕ defines a 1-dimensional space invariant under all the $f(g_j)$, and thus invariant for the representation of the group generated by the g_j. But since the representation is continuous, the space is also invariant for the representation of the closure of that group, which is all of $SO(3)$ (the closed subgroups of $SO(3)$ are known). But this would be a contradiction with the irreducibility of the representation f, unless it were the trivial representation. So the spectral radius of $f(g_1') + \cdots + f(g_8')$ cannot be 8, and the limit in (4.29) is 0, except for the trivial representation, proving that $\{g_n\}$ is uniformly distributed. $\square$

Basically, we achieved uniform distribution by using, in the substitution, rotation by $2\pi/3$ about the x axis ($R_x^{2\pi/3}$) and rotation by $2\pi/4$ about the y axis ($R_y^{2\pi/4}$). Repetition of the substitution gives rise to rotation by $R_y^{2\pi/4} R_x^{2\pi/3} R_y^{-2\pi/4}$, which is just $R_z^{2\pi/3}$ – that is, we have available to us rotation by $2\pi/3$ about both the x and z axes. (We will see the algebraic consequences of this in Theorem 4.7.) In the next example, the "dite & kart" tilings (made with Lorenzo Sadun

[RS1]), we achieve a similar effect with $2\pi/3$ replaced by $2\pi/5$. We give this new example to motivate the analysis of certain subgroups of $SO(3)$ which we call generalized dihedral groups, to show what is of *general* importance about the previous example.

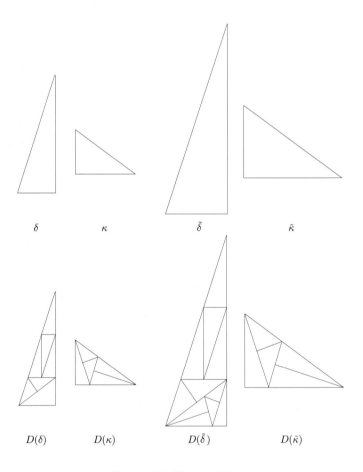

Figure 33. Dites and karts

The alphabet of the dite & kart tilings consists of 8 prisms. Consider the two right triangles of Fig. 33 denoted δ and κ. δ has legs of lengths 1 and $\tau\sqrt{2+\tau}$ and κ has legs of lengths τ and $\sqrt{2+\tau}/\tau$, where $\tau = (1 + \sqrt{5})/2$, the golden mean. (The triangles δ and κ are halves of the triangles S_A and S_B introduced by Raphael Robinson in his version of the kite & dart tilings [GrS; p. 540].) It is elementary to check that the small angle in δ is $\pi/10$ and the small angle in κ is $\pi/5$. We next introduce triangles $\tilde{\delta}$ and $\tilde{\kappa}$ which are larger than δ and κ by a linear factor τ.

Fig. 33 then shows how these 4 triangles can be decomposed into congruent copies of triangles which are each a linear factor τ^2 smaller than δ, κ, $\tilde{\delta}$ and $\tilde{\kappa}$.

We thicken δ by two different depths to make two types of prisms, a "short thin dite" of depth 1 and a "short thick dite" of depth τ. Likewise from $\tilde{\delta}$ we make a "*tall* thin dite" of depth 1 and a "*tall* thick dite" of depth τ. Finally, replacing δ by κ we make the analogous 4 types of "karts".

We have a similar procedure for the prisms, again shrinking by a linear factor of τ^2. We begin with the karts.

The short thin kart is decomposed into a pair of layers. The "top" layer consists of short thin dites and karts which, when viewed from above, have the same pattern as the decomposed 2-dimensional κ (Fig. 33). The bottom layer consists of short thick dites and karts in the same pattern. Since $\tau^2 = 1 + \tau$, the sum of the thicknesses of the two layers equals the thickness of the original thin kart.

The rule for the short *thick* kart is similar, only we now use 3 layers of short dites and karts; a top thin layer and 2 thick lower layers. Since $\tau = (1 + 2\tau)/\tau^2$, the total thickness of the decomposed layers equals the thickness of the original thick kart.

The rules for the *tall* thin and thick karts are now immediate, replacing the short dites and karts in the decomposition of $\tilde{\kappa}$ by tall dites and karts.

The rule for decomposed dites uses the decomposition of δ and $\tilde{\delta}$ rather than that of κ and $\tilde{\kappa}$, with one added twist, based on the rectangles appearing in the decomposition of δ and $\tilde{\delta}$ (Fig. 33). As with karts, the thin dites decompose into 2 layers and the thick ones into 3 layers. The decomposition of each dite generates 2 or 3 parallelepipeds corresponding to the aforementioned rectangle. If the original dite is short, then the parallelepiped in the thin layer has 2 square faces, while if the original dite is tall, the parallelepipeds in the thick layers have 2 square faces. We then rotate the parallelepiped by $\pi/2$ about the axis joining the centers of the square faces, as in Fig. 34. This completes the decomposition rules.

Finally, the dite & kart tilings are obtained by combining the above decomposition rules with an expansion about the origin by a linear factor of 2. The proof of the uniform distribution of the orientations of tiles in dite & kart tilings requires combining the use of Weyl's criterion, as we did for quaquaversal tilings, with the Perron-Frobenius theorem as we did for the pinwheel tilings.

We note that, as with the quaquaversal tilings, the uniform distribution is obtained by taking some rotation of finite order (now order 5, previously order 3) about the x axis and using a rotation by $\pi/2$ about an orthogonal axis (y) to produce rotation of order 5 about a third axis (z). As with the quaquaversal tilings, the angles of rotation that we use are "harmless" by themselves – not irrational as in the pinwheel – the uniform distribution now being caused by the *noncommutativity* of $R_x^{2\pi/3}$ and $R_z^{2\pi/3}$ (respectively $R_x^{2\pi/5}$ and $R_z^{2\pi/5}$). (More details about the quaquaversal and dite & kart tilings can be found in the original papers [CoR, RS1].)

We now want to squeeze some essential features from these 3-dimensional examples. We said earlier that what is new in three dimensions comes from using noncommutativity in place of irrationality. To understand just how, in these examples, the noncommutativity of the rotations replaces the irrationality of the pinwheel, it is natural to inquire into the *algebra* of such rotations. Specifically, we define

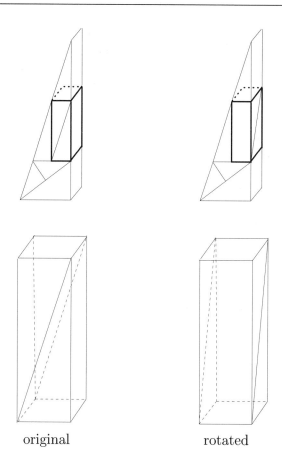

original   rotated

Figure 34. Rotating the boxes

$G(p, \ell, q)$ as the subgroup of $SO(3)$ generated by $R_x^{2\pi/p}$, $R_y^{2\pi/\ell}$ and $R_z^{2\pi/q}$, and abbreviate $G(p, 1, q)$ as $G(p, q)$. Note that $G(p, 2)$ is just the dihedral group of order $2p$, the group of symmetries of a regular p-gon in the plane. (In order to refer only to the plane the dihedral groups are usually thought of as generated by rotation in the plane by $2\pi/p$ and reflection in the plane about a line containing the center of rotation. From our point of view it is convenient to reinterpret these as the restriction to the plane of the action of 3-dimensional rotation by, respectively, $2\pi/p$ about an axis perpendicular to the plane, and rotation by π about a line in the plane intersecting the first line.) The quaquaversal and dite & kart tilings use $G(3,3)$ and $G(5,5)$, which leads to the natural desire to understand these groups $G(p,q)$ more generally. Basically, we will see that rotations about perpendicular axes are essentially independent of one another, except for some complications due to right angles. The (standard) notation used below is reviewed in Appendix II.

Theorem 4.7. *Let $p,q \geq 3$ and let $A \equiv R_x^{2\pi/p}$, $B \equiv R_z^{2\pi/q}$.*

1) If p and q are odd, then

$$G(p,q) = \langle A, B \ : \ A^p, B^q \rangle.$$

2) If p is even and q is odd, then

$$G(p,q) = \langle A, B \ : \ A^p, B^q, A^{p/2}BA^{p/2}B \rangle.$$

3) If p is even and q is $2 \times$ odd, then

$$G(p,q) = \langle A, B \ : \ A^p, B^q, A^{p/2}BA^{p/2}B, B^{q/2}AB^{q/2}A \rangle.$$

4) If p and q are divisible by 4, then

$$G(p,q) = G(\mathrm{lcm}(p,q), 1, 4) = G(\mathrm{lcm}(p,q), 1, 2) *_{G(4,1,2)} G(4,1,4),$$

where the obvious identifications are made in the amalgamation, and lcm *denotes least common multiple.*

Proof. We will only prove the cleanest part of the theorem, part 1), which shows that $G(p,q)$ is the free product of the cyclic groups generated respectively by A and B if and only if p and q are odd. The "only if" part is easy; if p, say, is even, then $A^{p/2}BA^{p/2}B = \mathbb{I}$ and

$G(p,q)$ is not the free product. To prove the other direction, what we need to prove is precisely that

$$A^{c_1} B^{d_1} A^{c_2} B^{d_2} \cdots \neq \mathbb{I}, \qquad (4.30)$$

if the c's (resp. d's) are nonzero modulo p (resp. q). Rather than work with a variety of rotations about two axes, it is convenient to reduce the variable behavior to just one axis together with appropriate use of a rotation by $\pi/2$; that is, we represent B and A in terms of $R_x^{2\pi/pq}$ and $R_y^{2\pi/4}$, by $A = [R_x^{2\pi/pq}]^q$ and $B = [R_y^{2\pi/4}]^3 [R_x^{2\pi/pq}]^p R_y^{2\pi/4}$. In terms of these rotations, (4.30) is a special case of the following lemma.

Basic Lemma 4.8. *Let p and q be odd, $m \equiv 4pq$, $T \equiv R_x^{2\pi/m}$ and $S \equiv R_y^{2\pi/4}$. Then:*

$$W S^{b_1} T^{a_1} S^{b_2} T^{a_2} \cdots S^{b_n} T^{a_n} E \neq \mathbb{I}, \qquad (4.31)$$

if $W, E \in G(4,4,1)$, $a_j \neq kpq$, b_j are odd, $n > 0$ and $\mathbb{I}$ is the identity matrix.

Proof of the lemma. Let $x = e^{2\pi i/m}$, $y = x^{pq} = e^{2\pi i/4}$ and $z = x^4 = e^{2\pi i/pq}$. Note that $y^4 = 1 = z^{pq}$. Let R be the ring $\mathbb{Z}[x] = \mathbb{Z}[y, z]$. Consider each factor $S^b T^a$ in the statement of the lemma. It is of the form

$$ST^a = \begin{pmatrix} 0 & -s & \tilde{c} \\ 0 & \tilde{c} & \tilde{s} \\ -1 & 0 & 0 \end{pmatrix} \qquad (4.32)$$

or

$$S^3 T^a = \begin{pmatrix} 0 & \tilde{s} & -\tilde{c} \\ 0 & \tilde{c} & \tilde{s} \\ 1 & 0 & 0 \end{pmatrix}, \qquad (4.33)$$

where $\tilde{c} = \cos(2\pi a/m) = (x^a + \bar{x}^a)/2$, and $\tilde{s} = \sin(2\pi a/m) = (x^a - \bar{x}^a)/2i$. Let I be a maximal extension of the ideal $(1 + y) \subset R$.

We need to prove:

Lemma 4.9. *If $a \neq kpq$, the $(1,2)$, $(1,3)$, $(2,2)$ and $(2,3)$ entries of the matrix $2S^b T^a$ are in R but not in the maximal ideal I.*

Proof. We prove this first for the $(2,2)$ entry $x^a + x^{-a} = y^u z^v + y^{-u} z^{-v}$, where we note that $v \neq 0$. Assume $y^u z^v + y^{-u} z^{-v} \in I$. Since $1 + y \in I$, $(-y)^u - 1 = -[1 + y][1 + (-y) + (-y)^2 + \cdots + (-y)^{u-1}] \in I$ and so $(-y)^u z^v - z^v \in I$. Similarly, $(-y)^{-u} - 1 \in I$, so $(-y)^{-u} z^{-v} - z^{-v} \in I$. This implies, using $y^u z^v + y^{-u} z^{-v} \in I$, that $z^v + z^{-v} \in I$. We now show that this implies $1 \in I$, which is a contradiction which proves the lemma for the $(2,2)$ entry.

Let $\tilde{z} \equiv z^v \neq 1$, and note that $\tilde{z}^{pq} = 1$. Now $(\tilde{z} + \tilde{z}^{-1})(\tilde{z}^2 + \tilde{z}^3) = (\tilde{z} + \tilde{z}^2 + \tilde{z}^3 + \tilde{z}^4) \in I$. Multiplying by $1 + \tilde{z}^4 + \tilde{z}^8 + \cdots + \tilde{z}^{4k}$ we see that $\tilde{z} + \tilde{z}^2 + \tilde{z}^3 + \cdots + \tilde{z}^{4k+4} \in I$. We now consider two cases. If $pq = 1 \pmod{4}$, take $k = (pq - 5)/4$, obtaining that $\tilde{z} + \tilde{z}^2 + \tilde{z}^3 + \cdots + \tilde{z}^{pq-1} \in I$. But $1 + \tilde{z} + \tilde{z}^2 + \tilde{z}^3 + \cdots + \tilde{z}^{pq-1} = (1 - \tilde{z}^{pq})/(1 - \tilde{z}) = 0$, so $\tilde{z} + \tilde{z}^2 + \tilde{z}^3 + \cdots + \tilde{z}^{pq-1} = -1$, which implies $1 \in I$. Alternatively, if $pq = 3 \pmod{4}$ take $k = (pq - 3)/4$, obtaining $\tilde{z} + \tilde{z}^2 + \tilde{z}^3 + \cdots + \tilde{z}^{pq+1} \in I$. But using $1 + \tilde{z} + \tilde{z}^2 + \tilde{z}^3 + \cdots + \tilde{z}^{pq-1} = (1 - \tilde{z}^{pq})/(1 - \tilde{z}) = 0$, we see that $\tilde{z} + \tilde{z}^2 + \tilde{z}^3 + \cdots + \tilde{z}^{pq+1} = \tilde{z}^{pq+1} = \tilde{z}$, and if $\tilde{z} \in I$ then $1 \in I$.

Now consider the other entries. The $(1,3)$ entry is just plus or minus the $(2,2)$ entry. The $(1,2)$ and $(2,3)$ entries are (up to sign) of the form $y^{u'} z^v + y^{-u'} z^{-v}$, where $u' = u + 1$. The above argument, with u replaced by u', shows that these elements are in R but not in I. This finishes the proof of the sublemma. $\qquad\square$

Getting back to the Basic Lemma, consider the matrix $2S^{b_i} T^{a_i}$. Note that $2 \in I$ since $2 = 1 - y^2 = (1 + y)(1 - y)$. We have shown that, modulo I, this matrix takes the form

$$\begin{pmatrix} 0 & \alpha & \beta \\ 0 & \gamma & \delta \\ 0 & 0 & 0 \end{pmatrix} \qquad (4.34)$$

with $\alpha, \beta, \gamma, \delta$ nonzero elements of the field R/I. But the product of two (or more) matrices of this form again takes this form, so $FS^{b_1}T^{a_1}S^{b_2}T^{a_2}\cdots S^{b_n}T^{a_n}$ (where F is a multiple of 2) again takes this form. Matrices in the group $G(4,4,1)$ are, up to sign, permutation matrices, so $FWS^{b_1}T^{a_1}S^{b_2}T^{a_2}\cdots S^{b_n}T^{a_n}E$ has 4 matrix elements that are nonzero in R/I. But F times the identity matrix is clearly zero modulo I, so $WS^{b_1}T^{a_1}S^{b_2}T^{a_2}\cdots S^{b_n}T^{a_n}E$ can never equal the identity.   □

This is a good time to tie together what we have covered in this chapter. We have been concerned with rotational properties of hierarchical tilings such as the kite & dart. (In this chapter we have downplayed the connection of this study of hierarchical tilings with the finite type tilings, which we emphasized earlier in the book; we are free to do this because Theorem 1.10, together with the explicit examples of the kite & dart and pinwheel, suggests that hierarchical tilings can always be thought of as finite type tilings. In fact there has recently been a rather general proof of this [Goo].) First we saw how the hierarchy leads to a *statistical form* of rotational symmetry. This was important historically as the underlying mechanism which, through diffraction, led to the discovery of quasicrystals, but it is also an important new mathematical idea. This feature is emphasized in those tilings, such as the 2-dimensional pinwheel and 3-dimensional quaquaversal, in which the symmetry group is large; and analysis of these tilings led to a classification of the generalized dihedral subgroups of $SO(3)$.

Chapter 5

Conclusion

I can watch a bubbling brook or waterfall for hours; the patterns are fascinating. If it were practical I would combine this curiosity with a scientific study of these patterns. Unfortunately the most interesting feature of such fluids, known as turbulence, is still beyond the understanding of science. The structural world of solids is pale by comparison – and indeed science has made much more progress in that direction. The (crystalline) structures of equilibrium solids are rather limited, and sophisticated mathematics has developed over many years to support such research, for instance the analysis of the crystallographic groups [HiC; p. 70]. The discovery of quasicrystals in 1984 [StO] reawakened interest in this old field, and coming as it did not long after the mathematical discovery of aperiodic tilings by Robert Berger, Raphael Robinson, Roger Penrose and others [GrS], there has been a commingling of the subjects.

Though not as rich as running water, the pinwheel and kite & dart tilings (Figs. 13 and 1) can still hold one's attention. All that structure built out of one or two simple shapes! Our analysis has emphasized two general features. First, they are global structures built out of many copies of a limited number of different components. And second, the complicated global structure results solely from the limited ways neighboring components interact. These are the main points. A third feature is perhaps serendipitous, less essential: the structures are hierarchical. Our main quest has been to discover what global structures can be produced by the local interactions of its small components, and in this sense the additional hierarchy is perhaps too special. But, as we saw in the last chapter, it has certainly proved fertile!

What have we learned after all from these pretty examples? We have seen at least three instances in which the study of these tilings has led to mathematics of independent interest, interest in a part of mathematics far from tiling. The first was the result of Shahar Mozes (Theorem 1.10) which tied together seemingly disparate specialties within ergodic theory – the subshifts of finite type and the substitution subshifts. Such unexpected connections are highly prized. The other two legacies of the study were the idea of statistical rotational symmetry, and the determination (Theorem 4.7) of the basic relations between simple pairs of rotations in space. Both of the latter are geometrical; in a sense, they are the result of working within the abstract framework of subshifts, but giving the abstract symbols a geometrical life.

The ideas we used to explore our vague "main quest" include ergodic theory, probability theory, statistical mechanics and ring theory. This was one of the goals of this book – to show that significant new mathematics can result from the interplay of widely disparate viewpoints.

As to our motivating problem on global structures, the picture we leave is incomplete; we have certainly not attained full understanding of such structures; all we have done is find a number of unexpected examples, and use them to obtain some unexpected consequences of the combination of hierarchy with rotations. There remain specific open problems, such as whether one can, by local rules, determine tilings all of which satisfy (1.1) – in which one loses all information traveling between far distant regions. And there are less concrete problems, such as exploring the relations between more complicated pairs of spatial rotations. There is much interesting work to be done, and I hope this book encourages further investigation. More generally, I hope I have given some sense of the value and excitement of analyzing an amorphous problem through a wide variety of the lenses available from mathematics and the physical sciences.

Appendix I. Geometry

We refer to the d-dimensional Euclidean spaces $\mathbb{R}^d$ only for $d = 1, 2, 3$. By a "congruence" or rigid motion we mean a map ϕ from such a space to itself which preserves the (Euclidean) distance between points: $d[\phi(x), \phi(y)] = d[x, y]$. The set of such maps constitutes the "Euclidean group" $\mathcal{E}^d$, which is generated by translations, rotations and reflections [NiS; p. 121]. Furthermore, the rotations can be identified (as a group) with $SO(d)$, the real $d \times d$ matrices of determinant 1 whose column vectors are orthonormal [Ree; p. 11].

A "symmetry" of a subset S of $\mathbb{R}^d$ is a congruence of $\mathbb{R}^d$ which maps S onto itself. More generally, a symmetry of a *collection* of subsets of $\mathbb{R}^d$ is a congruence of $\mathbb{R}^d$ which maps each set in the collection onto a set in the collection. The set of all symmetries of such a set (or collection of sets) is automatically a group under composition. For instance, the group of symmetries of a regular n-gon in $\mathbb{R}^2$ is the "dihedral group" of order $2n$, which is the group $\{a^p b^q : 0 \leq p \leq n - 1, \ 0 \leq q \leq 1\}$, where a is rotation of the n-gon about its center by $2\pi/n$ and b is reflection about a line joining some opposite pair of its vertices.

Consider a "tiling" T of $\mathbb{R}^d$ by polyhedra (that is, T is a collection of solid polyhedra, with pairwise disjoint interiors, whose union is

$\mathbb{R}^d$) for which the translational symmetries are of the form $n_1 t_1 + \cdots + n_d t_d$, for all $n_j \in \mathbb{Z}$, for d linearly independent vectors $t_j \in \mathbb{R}^d$. The possible symmetry groups for such tilings are called the "crystallographic groups", and have been classified [HiC, p. 81; NiS, p. 189].

Appendix II. Algebra

By a "representation" of the group G we mean a group homomorphism of G into the group of invertible linear functions of a complex vector space onto itself. The vector space is sometimes a space of functions, and sometimes has extra structure such as an inner product. A representation is called "n-dimensional" if the vector space is $\mathbb{C}^n$, and therefore the linear functions can be identified with $n \times n$ matrices with complex entries. It is then called "irreducible" if the only subspaces $M \subseteq \mathbb{C}^n$ mapped into themselves by all the matrices of the representation are $M = \{0\}$ and $M = \mathbb{C}^n$. For instance, the only irreducible representations of the circle group $G = SO(2)$ are one dimensional, and labeled by $\mathbb{Z}$, namely $f_m : \alpha \in [0, 2\pi) \to e^{im\alpha} \in \mathbb{C}$, $m \in \mathbb{Z}$; they are called "characters".

Given a finite group G, we define its "group algebra" as all formal complex linear combinations $\sum_{g \in G} a_g g$, where $a_g \in \mathbb{C}$. Note the natural one-to-one correspondence between the formal sum $\sum_{g \in G} a_g g$ and the complex-valued function on the group defined by $a(g) \equiv a_g$. We say two such formal sums are the same if and only if their associated functions are the same. Then we define multiplication by numbers, addition, and multiplication by, respectively:

- $\alpha \sum_{g\in G} a_g g \equiv \sum_{g\in G} (\alpha a_g) g$
- $(\sum_{g\in G} a_g g) + (\sum_{g'\in G} a'_g g) \equiv \sum_{g\in G} (a_g + a'_g) g$
- $(\sum_{g\in G} a_g g)(\sum_{g'\in G} a'_g g') \equiv \sum_{g,g'\in G} a_g a'_g g g'$

It is also useful to have an adjoint on this algebra: $(\sum_{g\in G} a(g)g)^* \equiv \sum_{g\in G} \bar{a}(g^{-1})g$.

Since we are representing groups as linear operators, we can extend the representation in a natural way to a representation of the group algebra. (Conversely, a representation of the algebra yields a unique representation of the group – but we won't need this.) And if the representation ϕ of the group is "unitary", that is, if $\phi(g)^* = [\phi(g)]^{-1} = \phi(g^{-1})$, then the representation of the algebra also respects the adjoint on the algebra: $\phi(a^*) = [\phi(a)]^*$ (see Appendix III).

Finally, we will use the notion of a "presentation" of a group. Let A be a nonempty collection of pairs of symbols a_j, a_j^{-1}. We define "words" to be finite ordered sequences of such symbols, such as $a_3 a_1^{-1} a_{17}$. The empty ordered sequence is denoted 1. A product is defined for such sequences by concatenation, for instance, the product $(a_3 a_1^{-1} a_{17})(a_{12} a_4) = a_3 a_1^{-1} a_{17} a_{12} a_4$. The symbol a^n is an abbreviation for the word $aaa\ldots a$ consisting of n a's. Given a word $W = a_{j_1} a_{j_2} \ldots a_{j_k}$, the word W^{-1} is defined to be $W^{-1} = a_{j_k}^{-1} \ldots a_{j_2}^{-1} a_{j_1}^{-1}$.

Assume given, besides A, some set (possibly empty) of words $P_1, P_2 \ldots$. We define an equivalence relation on the set of words: $W_1 \sim W_2$ if W_1 can be turned into W_2 by a finite number of applications of the following operations:

- Insertion of one of the words $a_j a_j^{-1}$, $a^{-1} a_j$, P_j, or P_j^{-1}, either between two consecutive symbols in W_1, or just before the first symbol of W_1 or just after the last symbol of W_1.

- Deletion of one of the words $a_j a_j^{-1}$, $a_j^{-1} a_j$, P_j, or P_j^{-1}, if it forms a consecutive block of symbols in W_1.

We denote the set of equivalence classes of words by $\langle a_1, a_2, \dots :$ $P_1, P_2, \dots \rangle$. It is easily seen to be a group with product defined by concatenation. In this notation the dihedral group of order $2n$ is $\langle a_1, a_2 : a_1^n, a_2^2, a_1 a_2 a_1 a_2 \rangle$. For another example consider the two finite cyclic groups $\mathbb{Z}_3$, $\mathbb{Z}_5$ of orders 3 and 5 respectively. Their "free product", denoted $\mathbb{Z}_3 * \mathbb{Z}_5$, is the group $\langle a_1, a_2 : a_1^3, a_2^5 \rangle$; that is, all finite strings of a_1's and a_2's of appropriate powers.

There is a generalization of free products which is of significant interest. Consider the group G with presentation $\langle a, b : a^4, b^6, a^2 b^{-3} \rangle$. This is almost the free product of $\mathbb{Z}_4 = \langle a : a_4 \rangle$ with $\mathbb{Z}_6 = \langle b : b^6 \rangle$. Both these groups have $\mathbb{Z}_2$ as a subgroup: the subgroups generated by a^2 and b^3, respectively; and G is said to be the *free product* of its subgroups generated by a and b, respectively, *amalgamated* over the common subgroup ($\mathbb{Z}_2$) – that is, with this "common" subgroup identified. (Think of this as an algebraic version of the process in topology wherein you "identify" the opposite edges of a square to get a torus. In fact, there are topological issues here, but we will ignore them as they are tangential to our subject.) This is expressed: $G = \mathbb{Z}_4 *_{\mathbb{Z}_2} \mathbb{Z}_6$, though this notation does not show the way in which the common subgroups of the factors are identified. More generally, the group $G \equiv$ $\langle a_1, a_2, \dots, b_1, b_2, \dots : R(a_1), \dots, S(b_1), \dots, U_1(a_{j_1})[V_1(b_{k_1})]^{-1}, \dots \rangle$ is called the "amalgamated free product" of the subgroup A generated by the a's and the subgroup B generated by the b's, with the subgroup generated by $U_1(a_{j_1}), \dots$ of A amalgamated with the subgroup generated by $V_1(b_{k_1}), \dots$ of B. If this common subgroup is H, we write $G = A *_H B$. For an elaboration of the above we refer to the classic text [MKS], and for other aspects of introductory algebra we recommend [Her].

Appendix III. Analysis

The first topic in this appendix is abstract integration; for further details we recommend the text [Tay]. Let X be a compact metrizable space, and let $C(X)$ be the complex continuous functions on X. We wish to "integrate" functions on X, including those in $C(X)$. For instance, let X be a countable product of 2-point spaces, $X = \{0,1\}^{\mathbb{Z}} = \{x_j : x_j \in \{0,1\}, j \in \mathbb{Z}\}$. By Tychonoff's theorem X is compact if we take the topology for the product to be generated by so-called "cylinder sets", that is, sets of the form $X_\epsilon \equiv X_{\epsilon_{j_1}, \ldots \epsilon_{j_k}} \equiv \{x_j : x_{j_1} = \epsilon_{j_1}, \ldots, x_{j_k} = \epsilon_{j_k}\}$, where $\epsilon_j \in \{0,1\}$. Indicator functions for such cylinder sets (that is, functions which have value 1 on the set and value 0 off the set) are continuous. Now when X is used as the model for the history of flips of a fair coin [Bil], it is natural to associate with the cylinder set $X_{\epsilon_{j_1}, \ldots \epsilon_{j_k}}$ an "expected value" $(1/2)^k$, which would be the value of the "integral" $\mathbb{I}(\chi_{X_\epsilon})$ of the indicator function χ_{X_ϵ}.

In general, an expected value or (probability) integral $\mathbb{I}$ would have the properties:

- $\mathbb{I}(f) \geq 0$ for any $f \in C(X)$, $f \geq 0$
- $\mathbb{I}(I) = 1$ (where I is the constant function on X with value 1)
- $\mathbb{I}(af + bg) = a\mathbb{I}(f) + b\mathbb{I}(g)$, for $a, b \in \mathbb{C}$, $f, g \in C(X)$ – that is, $\mathbb{I}$ is linear

Starting with any such integral one can uniquely extend its domain to a larger class of "integrable" functions, denoted $\mathcal{L}_1^{\mathbb{I}}$. For instance, consider the class of "Borel" subsets S of X, that is, those sets which can be made from open sets by countably many operations of union, intersection and complement. (Any countable set is easily seen to be Borel.) The indicator function χ_S of a Borel set S is always integrable. Similarly, for a positive integer p one defines $\mathcal{L}_p^{\mathbb{I}}$ as the set of functions f for which f^p is integrable. These spaces usually include some nonzero functions f for which $\mathbb{I}(|f|) = 0$; call this set Z. It is an important theorem of Markov that when we use Z to define equivalence classes in $\mathcal{L}_p^{\mathbb{I}}$ (two functions being equivalent if their difference is in Z), the quotient, denoted $L_p^{\mathbb{I}}$, is a linear space which is *complete* in the norm $\|\{f\}\|_p^{\mathbb{I}} \equiv [\mathbb{I}(|f|^p)]^{1/p}$, where f is any representative of the equivalence class $\{f\} \in L_p^{\mathbb{I}}$. In fact, $L_p^{\mathbb{I}}$ can be identified with the completion of $C(X)$ in the norm $\| \cdot \|_p^{\mathbb{I}}$. Also, the linear space $L_2^{\mathbb{I}}$, on which we can put an inner product $\langle \{f\}, \{g\} \rangle_{\mathbb{I}} \equiv \mathbb{I}(\bar{f}g)$, is thus a "Hilbert space": that is, it is complete in the norm coming from its inner product $(\|\{f\}\|_2^{\mathbb{I}} = [\langle \{f\}, \{f\} \rangle_{\mathbb{I}}]^{1/2})$. (The usual inner product on $\mathbb{C}^d$ is given by $\langle x, y \rangle = \sum_j \bar{x_j} y_j$.) A simple technical result about vectors in a Hilbert space which has wide applicability is the Cauchy-Schwarz inequality: for all f, g,

$$|\langle f, g \rangle|^2 \leq \langle f, f \rangle \langle g, g \rangle. \qquad (III.1)$$

Another important technical result we will use is the

Monotone Convergence Theorem. *Suppose f_n is a sequence of real functions on X, $f_n \in \mathcal{L}_1^{\mathbb{I}}$, satisfying $f_{n+1} \geq f_n$. Assume $\lim_n f_n(x) = f(x)$ for all $x \in X$. Then $f \in \mathcal{L}_1^{\mathbb{I}}$ if and only if $\lim_n \mathbb{I}(f_n) < \infty$. When this is the case, $\lim_n \mathbb{I}(f_n) = \mathbb{I}(f)$.*

The other main topic of this appendix is operator theory; we recommend [BaN, RiN, Nai] as general references. First we classify some of the operators on a (finite or infinite dimensional) Hilbert space $\mathcal{H}$. The continuous linear operators, that is, the set of all maps $A : \mathcal{H} \longrightarrow \mathcal{H}$ which are linear and continuous, is denoted $B(\mathcal{H})$.

Appendix III. Analysis

The "adjoint" $A^* \in B(\mathcal{H})$ of $A \in B(\mathcal{H})$ is that operator such that $\langle x, Ay \rangle = \langle A^*x, y \rangle$ for all $x, y \in \mathcal{H}$. (Thinking of $d \times d$ matrices as operators on the inner product space $\mathbb{C}^d$ in the usual way, the components of the adjoint M^* are related to those of M by $M_{j,k}^* = \overline{M_{k,j}}$.) We will be referring to three subsets of $B(\mathcal{H})$: the "self-adjoint" operators A are those equal to their adjoints, that is, $A = A^*$; the "projections" P are those self-adjoint operators which are idempotent, that is, $P^2 = P$; and the "unitary" operators U are those which preserve the inner product, namely, $\langle Ux, Uy \rangle = \langle x, y \rangle$ for all $x, y \in \mathcal{H}$. The range of a projection P is a closed linear subspace of $\mathcal{H}$, and every such subspace is the range of a projection. The unitaries preserve all the structure of the Hilbert space, so they constitute its "symmetries".

There are numerous convenient ways to measure the size of an operator $A \in B(\mathcal{H})$. One is its norm $\|A\| \equiv \sup_{x \in \mathcal{H}: \|x\|=1} \|Ax\|$. Another is its "spectral radius" $\sigma(A) \equiv \sup\{|\mu| : \mu \in \mathbb{C}, (\mu - A)^{-1} \notin B(\mathcal{H})\}$. The two are related by $\sigma(A) = \lim_{n \to \infty} \|A^n\|^{1/n}$.

Our main use of Hilbert space is to house representations of groups, in particular the discrete group $\mathbb{Z}^d$ and the continuous group $\mathbb{R}^d$. They are usually represented by unitary operators T^g on some Hilbert space $\mathcal{H}$, with the representation "continuous" in the sense that $T^g y$ is continuous in $g \in G$ for each fixed $y \in \mathcal{H}$ (or equivalently, that $\langle x, T^g y \rangle$ is continuous in $g \in G$ for fixed $x, y \in \mathcal{H}$).

References

[BaN]   G. Bachman and L. Narici, *Functional Analysis*, Academic Press, New York, 1966.

[Bil]   P. Billingsley, *Ergodic Theory and Information*, John Wiley, New York, 1965.

[CoR]   J.H. Conway and C. Radin, *Quaquaversal tilings and rotations*, Inventiones Math. **132** (1998), 179–188.

[Cul]   K. Culik II, *An aperiodic set of* 13 *Wang tiles,*, Discrete Appl. Math. **160** (1996), 245–251.

[DeK]   F. M. Dekking and M. Keane, *Mixing properties of substitutions,*, Z. Wahrsch. Verw. Gebiete **42** (1978), 23–33.

[Dwo]   S. Dworkin, *Spectral theory and x-ray diffraction*, J. Math. Phys. **34** (1993), 2965–2967.

[EiI]   A. Einstein and L. Infeld, *The Evolution of Physics: The Growth of Ideas from Early Concepts to Relativity and Quanta*, Simon and Schuster, New York, 1938.

[Gar]   M. Gardner, *Extraordinary nonperiodic tiling that enriches the theory of tiles*, Sci. Amer. (USA) (December 1977), 110–119.

[Goo]   C. Goodman-Strauss, *Matching rules and substitution tilings*, Ann. of Math. (2) **147** (1998), 181–223.

[GrS]   B. Grünbaum and G.C. Shephard, *Tilings and Patterns*, Freeman, New York, 1987.

[Gui]   A. Guinier (translated by P. Lorrain and D. Ste.-M. Lorrain), *X-ray Diffraction in Crystals, Imperfect Crystals, and Amorphous Bodies*, Freeman, San Francisco, 1963.

[Her]   I.N. Herstein, *Topics in Algebra*, Blaisdell, New York, 1964.

[HiC]   D. Hilbert and S. Cohn-Vossen, *Geometry and the Imagination*, Chelsea, New York, 1952.

[Kol]   A.N. Kolmogorov (translated by N. Morrison), *Foundations of the Theory of Probability*, 2nd English ed., Chelsea, New York, 1956.

[Kra]   H.A. Kramers, Remark in the discussion of P.W. Bridgman's paper, *New vistas for intelligence*, Physical Science and Human Values: A Symposium (E.P. Wigner, ed.), Princeton University Press, Princeton, NJ, 1947, pp. 156–157.

[KuN]   L. Kuipers and H. Niederreiter, *Uniform Distribution of Sequences*, Wiley, New York, 1974.

[MKS]   W. Magnus, A. Karrass and D. Solitar, *Combinatorial Group Theory: Presentations of Groups in Terms of Generators and Relations*, 2nd ed., Dover, New York, 1976.

[Moz]   S. Mozes, *Tilings, substitution systems and dynamical systems generated by them*, J. Analyse Math. **53** (1989), 139–186.

[Nai]   M.A. Naimark (translated by L. Boron), *Normed Rings*, Noordhoff, Groningen, 1964.

[NiS]   V.V. Nikulin and I.R. Shafarevich, *Geometries and Groups*, Springer-Verlag, Berlin, 1994.

[Pet]   K. Petersen, *Ergodic theory*, Cambridge University Press, Cambridge, 1983.

[Ra1]   C. Radin, *Low temperature and the origin of crystalline symmetry*, Internat. J. Modern Phys. **B1** (1987), 1157–1191.

[Ra2]   C. Radin, *The pinwheel tilings of the plane*, Ann. of Math. (2) **139** (1994), 661–702.

[Ra3]   C. Radin, *Space tilings and substitutions*, Geometriae Dedicata **55** (1995), 257–264.

[Ra4]   C. Radin, *Symmetry and tilings*, Notices Amer. Math. Soc. **42** (1995), 26–31.

[RaW]   C. Radin and M. Wolff, *Space tilings and local isomorphism*, Geometriae Dedicata **42** (1992), 355–360.

[Ree]   E.G. Rees, *Notes on Geometry*, Springer-Verlag, Berlin, 1983.

[RiN]   F. Riesz and B. Sz-Nagy (translated by L. Boron), *Functional Analysis*, Ungar, New York, 1955.

[Rob]   R.M. Robinson, *Undecidability and nonperiodicity for tilings of the plane*, Invent. Math. **12** (1971), 177–209.

[RS1]   C. Radin and L. Sadun, *Subgroups of* SO(3) *associated with tilings*, J. Algebra **202** (1998), 611–633.

[RS2]   C. Radin and L. Sadun, *On 2-generator subgroups of* SO(3), Trans. Amer. Math. Soc., to appear.

[Rue]   D. Ruelle, *Statistical Mechanics: Rigorous Results*, Benjamin, New York, 1969.

[Sen]   M. Senechal, *Quasicrystals and Geometry*, Cambridge University Press, Cambridge, 1995.

[StO]   P. Steinhardt and S. Ostlund, *The Physics of Quasicrystals*, World Scientific, Singapore, 1987.

[Tay]   A. Taylor, *General Theory of Functions and Integration*, Blaisdell, Waltham, MA, 1965.

[Wal]   P. Walters, *An Introduction to Ergodic Theory*, Springer-Verlag, New York, 1982.

[Wan]   H. Wang, *Games, logic and computers*, Sci. Amer. (USA) (November 1965), 98–106.

List of Symbols

Subject Index